Lecture Notes in Mathematics

Edited by A. Dold and B. Eckmann

1175

Karl Mathiak

Valuations of Skew Fields and Projective Hjelmslev Spaces

Springer-Verlag
Berlin Heidelberg New York Tokyo

Author

Karl Mathiak
Institut für Algebra und Zahlentheorie
Technische Universität Braunschweig
Pockelsstr. 14, 3300 Braunschweig, Federal Republic of Germany

Mathematics Subject Classification (1980): 10F45, 12E15, 12J10, 16A05, 16A10, 16A39, 51C05

ISBN 3-540-16099-X Springer-Verlag Berlin Heidelberg New York Tokyo
ISBN 0-387-16099-X Springer-Verlag New York Heidelberg Berlin Tokyo

Printed in Germany

Printing and binding: Beltz Offsetdruck, Hemsbach/Bergstr.
2146/3140-543210

Preface

This text is intended to provide a reasonably self-contained account of the theory of valuations of skew fields and their application to projective Hjelmslev spaces. The reader is only assumed to be familiar with the basic notions of algebra and the rudiments of topology and geometry. Nevertheless, the acquaintance with Krull valuations would be helpful (see for instance Endler [1] or Krull [1]).

The concept of valuation used here is due to F. Radó. It is more general than that of a Schilling valuation which is characterized by the fact that the valuation ring of the valuation is invariant under the inner automorphisms of the field. In contrast to this, the valuation rings of general valuations may be invariant or not.

The principal difference between general valuations and invariant valuations is the following: The value set of a general valuation is only a totally ordered set and has no algebraic structure. It contains therefore less information about the ideal structure of the valuation ring than the value set of a Krull valuation. To overcome this disadvantage, the concept of a value group is introduced which is a subgroup of the automorphism group of the value set. Using a slightly modified notion of convexity in the value group, Krull's theorems on the correspondence between prime ideals and convex subgroups may be generalized to skew fields.

If a valuation is invariant, the value group can be embedded into the value set. Then the algebraic structure of the value group is carried over onto the value set. This leads back to the concept of a Krull valuation where the non-zero values form a group.

There are two methods to construct fields with non-invariant valuations. These fields are formal power series fields and skew rational function fields. The existence of non-invariant valuations on these fields follows from two extension theorems. It should be mentioned that no general extension theory for valuations of skew fields exists.

The definition of valuation is asymmetric with respect to multiplication in the field. Considering valuation topologies naturally leads to the notion of V-valuations. From an algebraic point of view they have a higher degree of symmetry than valuations in general. Little is known about valuations which are not V-valuations. They behave rather strangely with respect to their algebraic and topological properties. This will become clear in the discussion of the approximation theorem, in particular in the discussion of the independence of valuations.

The theory of valued vector spaces surely deserves attention for its own sake. Here, this theory is only treated in view of its geometric applications. A set of projective Hjelmslev spaces is attached to each valued vector space. The main result is the description of the union and intersection sets in these spaces. This description yields the dimension formula which is one of the fundamental properties of these spaces. This formula is no longer valid in projective Hjelmslev spaces over local rings.

Hjelmslev's intention [1] was the construction of natural geometries for which uniqueness of join and intersection does not hold in general. For instance, when two lines form a small angle (schleifender Schnitt) their intersection is not unique. Hjelmslev assumes uniqueness only in the case of orthogonal lines.

In his concept of a projective geometry with homomorphism, Klingenberg [3] avoids metric notions. He claims that the intersection of two lines is unique if and only if they are mapped onto distinct elements by the given homomorphism. We shall introduce a similar criterion of uniqueness assigning to the ideal I a certain completely prime ideal $P_r(I)$ resp. $P_l(I)$. If I itself is completely prime, the criterion implies that H_I is an ordinary projective space.

In an appendix Graeter gives an account of the theory of locally invariant valuations which are special V-valuations. He succeeds in generalizing Ribenboim's approximation theorem where the compatibility conditions are formulated by means of the least upper bound of two valuations. Moreover, he shows that this theorem cannot be extended to a greater class of valuations.

Each chapter ends with a collection of exercises some of which serve to present supplementary results. The solutions are given at the end of the text.

CONTENTS

Terminology

Throughout, K denotes a not necessarily commutative field, $K^{\times}$ the multiplicative group of K. The prefix "skew" is only used for emphasis.

If W is the value set of a valuation, $W^{\times}$ denotes the set of the non-zero values of W.

A ring R with a unit element is said to be **local**, if the set of the non-invertible elements is an ideal M of R. The factor ring R/M is a field and is called **residue class field**.

We shall deal almost exclusively with left vector spaces and left modules and refer to these simply as vector spaces and modules.

A two-sided ideal P of a ring R is said to be **completely prime** if $P \neq R$ and if the factor ring R/P has no zero divisors.

An element c of a partially ordered set H is said to be an upper bound of two elements $a, b \in H$ if $a, b \leq c$. The set V(a, b) of all minimal upper bounds is called the union set of a and b. Analogously, the intersection set S(a, b) is defined as the set of all maximal lower bounds of a and b. If H is a lattice, then V(a, b) resp. S(a, b) contains exactly one element.

If H is a partially ordered set and if $a \leq b$ or $b \leq a$ holds for any pair of elements, then H is said to be **totally ordered**. A subset S of H is **convex** if $a \leq b \leq c$ and $a, c \in S$ imply $b \in S$. Further, S is called an **upper class** if $a \leq b$ and $a \in S$ imply $b \in S$. A lower class is similarly defined.

A **filter** τ on a set S is a family of subsets of S such that

(1) the intersection of two members of τ belongs to τ,

(2) any subset of S containing a member of τ belongs to τ.

A **filter base** is a family of subsets of S such that the intersection of finitely many members always contains an element of the filter base. When a filter base is given, then the family of all subsets of S containing a member of the filter base is a filter on S. This filter is said to be generated by the filter base.

CHAPTER 1 VALUATIONS OF FIELDS

In this chapter we present the basic definitions of the valuation theory of skew fields. After some preliminary results we discuss in more detail the ideal structure of valuation rings using the value group of the valuation. Finally, we shall describe two methods to construct fields with non-invariant valuations.

1. Valuation rings

A subring B of a field K is called a **valuation ring** if every element of K is either in B or is the inverse of an element of B.

It should be mentioned that this definition of a valuation ring differs from that of Schilling who requires further $a^{-1}Ba = B$ for all $a \in K^{\times}$. Radó uses the term total subring.

A non-empty subset I of K is said to be a fractional right ideal of K with respect to a valuation ring B if $Ia \subseteq I$ for all $a \in B$. Similarly fractional left ideals are defined. When I is contained in B then I is called integral. Examples of fractional right ideals are the principal right ideals aB for $a \in K$, Obviously, aB is integral if and only if $a \in B$.

The following shows that the integral fractional right ideals are right ideals in the ring theoretical sense.

PROPOSITION 1.1. The fractional right ideals are right B-submodules of K.

PROOF: We must only show that a fractional right ideal I is a subgroup of K with respect to addition. Let a and b be non-zero elements of I. It follows immediately from the definition of a valuation ring that $1 \in B$ and that $a^{-1}b \in B$ or $b^{-1}a \in B$. Therefore we obtain

$$a - b = a(1 - a^{-1}b) = b(b^{-1}a - 1) \in IB \subseteq I.$$

Similarly it is shown that a fractional left ideal is a left B-submodule of K.

THEOREM 1.2. The set of all fractional right (resp. left) ideals is totally ordered by inclusion.

PROOF: Let I and J be two fractional right ideals. Suppose that I is not contained in J. Choose an element $a \in I$ which is not in J. In order to show $J \subseteq I$ let $b \in J$, $b \neq 0$. Since B is a valuation ring, $b^{-1}a \in B$ or $a^{-1}b \in B$. If $b^{-1}a \in B$ then $a = b(b^{-1}a) \in J$ contrary to the choice of a. Hence $a^{-1}b \in B$ and therefore $b = a(a^{-1}b) \in I$. The assertion for fractional left ideals is similarly proved.

Let B be a valuation ring and **O** the family of all subrings of K containing B. Since for $B' \in \mathbf{O}$ $a \notin B'$ implies $a \notin B$, hence $a^{-1} \in B \subseteq B'$, every $B' \in \mathbf{O}$ is a valuation ring of K. Further, B′ is fractional right ideal with respect to B because $B'B \subseteq B'B' \subseteq B'$. We therefore obtain

COROLLARY 1.3. The set **O** of all subrings of K containing B is totally ordered by inclusion.

By using the inverses in the field we can define a bijective mapping from the chain of the fractional right ideals to the chain of the fractional left ideals. Let I be a non-empty subset of K. We define

$$I^* = \{ a \in K \mid a = 0 \text{ or } a^{-1} \notin I \}$$

or equivalently

$$I^* = \{ a \in K \mid 1 \notin Ia \} = \{ a \in K \mid 1 \notin aI \}.$$

Obviously, I^* is a fractional left ideal when I is a fractional right ideal and conversely.

PROPOSITION 1.4. The mapping $I \mapsto I^*$ is a one-to-one antitone correspondence of both ideal chains. $I^{**} = I$ holds for any fractional right or left ideal I.

PROOF: From the definition we conclude immediately that the star-mapping is antitone. Let $a \in K^{\times}$. Then we have

$$a \in I^{**} \Leftrightarrow a^{-1} \notin I^* \Leftrightarrow 1 \in Ia^{-1} \Leftrightarrow a \in I,$$

hence $I^{**} = I$. This implies that $I \mapsto I^*$ is bijective.

PROPOSITION 1.5. A valuation ring is a local ring. If M denotes the maximal ideal of B we have $B^* = M$ and $M^* = B$.

PROOF: Let M be the set of all non-units of B. Then $a \in M$, $a \neq 0$, implies $a^{-1} \notin B$, hence $a \in B^*$ and therefore $M \subseteq B^*$. Conversely, let $a \in B^*$, $a \neq 0$. From the definition of the star-mapping we conclude $a^{-1} \notin B$, hence a is a non-unit. This shows $B^* = M$. Thus M is an ideal of B. This means that B is a local ring. Finally, $M^* = B^{**} = B$ by 1.4.

PROPOSITION 1.6. Let B be a valuation ring and M its maximal ideal. Then the following conditions are equivalent

(1) P is a completely prime ideal of B.

(2) P is a non-empty subset of M satisfying the condition

(P) $a, b \in K$ and $ab \in P$ imply $a \in P$ or $b \in P$.

(3) P is a right ideal of B and P^* is a valuation ring containing B.

(4) P is the maximal ideal of a valuation ring B' containing B.

PROOF: (1) ⇒ (2). For $a, b \in B$ the condition (P) is satisfied since B/P has no zero-divisors. On the other hand, if a or b are not in B then $a^{-1} \in B$ or $b^{-1} \in B$. Since P is a two-sided ideal we have $b = a^{-1}(ab) \in P$ or $a = (ab)b^{-1} \in P$.

(2) ⇒ (3). First we show that P is a right ideal. Let $a \in P$, $b \in B$. Suppose ab is not in P. Then $a = (ab)b^{-1} \in P$ implies $b^{-1} \in P \subseteq M$ by (P), hence $1 = bb^{-1} \in M$, a contradiction. Thus $ab \in P$ and P is a right ideal of B. (By 1.1 we need not prove that P is a group with respect to addition.)

Since $P \subseteq M$ we have $B = M^* \subseteq P^*$. Further, P^* is multiplicatively closed. For, let a and b be non-zero elements of P^*. If ab is not in P^* then $(ab)^{-1} = b^{-1}a^{-1} \in P$. By (P) we have $a^{-1} \in P$ or $b^{-1} \in P$, hence a or b are not in P^* contrary to the choice of a and b. Thus P^* is a ring containing B.

(3) ⇒ (4). Setting $B' = P^*$ we have $B'^* = P^{**} = P$ which is the maximal ideal of B' by 1.5.

(4) ⇒ (1). We have $P = B'^* \subseteq B^* = M$, hence P is a proper ideal of B. Since P is the maximal ideal of B', the factor ring B'/P and hence B/P has no zero-divisors. Thus P is a completely prime ideal of B.

By 1.6 the star-mapping yields a one-to-one antitone correspondence of the set of all completely prime ideals of B and the set of all subrings of K containing B. Therefore we have

THEOREM 1.7. Let $\mathfrak{P}$ denote the set of all completely prime ideals of B and $\mathbf{O}$ the set of all subrings of K containing B. Then the mapping

$$\mathfrak{P} \longrightarrow \mathbf{O}, \quad P \longmapsto P^*$$

is an antitone bijection.

The inverse mapping $\mathbf{O} \longrightarrow \mathfrak{P}$ is also the star-mapping. If $B' \in \mathbf{O}$ then B'^* is the maximal ideal of B' and simultaneously a completely prime ideal of B.

2. Valuations

Let K be a field and W a totally ordered set with a least element denoted by 0. A **valuation** of K is a surjective mapping $K \longrightarrow W$, $a \longmapsto |a|$, satisfying

(B1) $|a| = 0 \Leftrightarrow a = 0$.

(B2) $|a + b| \leq \mathrm{Max}(|a|, |b|)$.

(B3) For every $a \in K$ there exists an order preserving mapping $\tilde{a}\colon W \longrightarrow W$ such that $\tilde{a}|b| = |ab|$ for all $b \in K$.

The main difference between this concept and a Schilling valuation (or a Krull valuation in the commutative case) is that the value set W has no algebraic structure. A valuation only allows the comparison of the elements according to their values in W.

PROPOSITION 2.1. Let $|\ |\colon K \longrightarrow W$ be a valuation of K. Then $|a| = |-a|$ for all $a \in K$.

PROOF: Since W is totally ordered, one of the inequalities $|a| \leq |-a|$ or $|-a| \leq |a|$ holds. By applying the order preserving mapping associated with -1 one inequality turns into the other, hence $|a| = |-a|$.

PROPOSITION 2.2. If $|a| < |b|$ then $|a + b| = |b|$.

PROOF: By (B2) $|a + b| \leq |b|$. From $|a + b| < |b|$ it follows that

$$|b| = |(a + b) - a| \leq \mathrm{Max}(|a + b|, |-a|) < |b|,$$

a contradiction, hence $|a + b| = |b|$.

This proposition is sometimes called the **principle of domination**. It shall be used extensively in the following.

THEOREM 2.3. 1.) Let $|\ |\colon K \longrightarrow W$ be a valuation of K. Then

$$B = \{\ a \in K \mid |a| \leq |1|\ \}$$

is a valuation ring of K and $M = \{\ a \in K \mid |a| < |1|\}$ its maximal ideal.

2.) Let B be a valuation ring of K. Then there exists a valuation of K with B as associated valuation ring.

PROOF: 1.) By (B2) and 2.1 B is a group with respect to addition. Further, if a, b $\in$ B then

$$|ab| = \tilde{a}|b| \leq \tilde{a}|1| = |a| \leq |1|,$$

hence ab $\in$ B. Thus B is a subring of K.

Let a $\in$ K. Since the value set W is totally ordered $|a| \leq |1|$ or $|1| \leq |a|$ holds. In the first case we have a $\in$ B, in the second case $|a^{-1}| \leq |1|$, hence a is the inverse of an element of B. Thus B is a valuation ring.

If $|a| = |1|$ then $|a^{-1}| = |1|$, too, hence a is a unit in B. Conversely, if a is a unit in B then $|a| \leq |1|$ and $|a^{-1}| \leq |1|$, hence $|a| = |1|$. Therefore, a is a unit in B if and only if $|a| = |1|$. Since M is the set of the non-units of B we have $M = \{ a \in K \mid |a| < |1| \}$.

2.) Let B be a valuation ring of K and $W = \{ aB \mid a \in K \}$ the set of all fractional principal right ideals. 1.2 shows that W is totally ordered by inclusion. The null ideal is the least element of W.

Setting $|a| = aB$ we obtain a surjective mapping $| \ | : K \longrightarrow W$. Clearly $|a| = aB = 0$ if and only if $a = 0$. Since W is totally ordered we have

$$(a + b)B \subseteq aB + bB = \mathrm{Max}(aB, bB)$$

hence

$$|a + b| \leq \mathrm{Max}(|a|, |b|).$$

For a $\in$ K we define $\tilde{a} : W \longrightarrow W$, $bB \longmapsto abB$. This mapping preserves the order of W and $\tilde{a}|b| = |ab|$ is valid for all b $\in$ K. Further $\{ a \in K \mid aB \subseteq B \} = B$. Hence the associated valuation ring is B.

Two valuations with the same valuation ring are said to be equivalent. Obviously, $| \ |_1$ and $| \ |_2$ are equivalent if and only if

$$|a|_1 \leq |b|_1 \iff |a|_2 \leq |b|_2$$

for all a, b $\in$ K. By 2.3 there is a one-to-one correspondence between valuation rings and equivalence classes of valuations. In defining valuations the goal has been that every valuation ring of the field corresponds to a valuation.

An endomorphism of W is a mapping preserving the order of W. The set of all endomorphisms of W is denoted by End W.

Let $|\ |: K \longrightarrow W$ be a valuation of K. By (B3) there exists a mapping

$$v: K \longrightarrow \mathrm{End}\, W, \quad v(a) = \tilde{a}.$$

We have

$$v(ab)|c| = |abc| = v(a)v(b)|c|$$

for all $c \in K$. Since a valuation is a surjective function this implies

$$v(ab) = v(a)v(b).$$

Further, $v(0)$ is the null function of W and $v(1) = \mathrm{id}$ is the identity on W. For $a \neq 0$, $v(a)$ is an automorphism because

$$v(a)v(a^{-1}) = v(a^{-1})v(a) = v(1) = \mathrm{id}.$$

The image

$$G = v(K^{\times}) = \{ v(a) \mid a \in K^{\times} \}$$

is a group called **value group** of the valuation. Observe that the value group is not a subset of the value set W, but a subgroup of the automorphism group of W.

We define

$$\varphi: G \longrightarrow W, \ \varphi(g) = g|1|$$

and call φ the **canonical mapping** from G to W. If $g = v(a)$ then $\varphi(g) = |a|$.

We may consider G as a group operating on W. The stabilizer of $|1|$ is a subgroup of G which we denote by U_0. We have

$$U_0 = \{ v(a) \in G \mid |a| = |1| \},$$

hence U_0 is the image of the group of units of B under the homomorphism v.

PROPOSITION 2.4. The mapping

$$a^{-1}Ba \longmapsto \tilde{a}^{-1}U_0\tilde{a}, \quad a \neq 0,$$

is a bijection of the set of conjugates of B onto the set of conjugates of U_0.

PROOF : 1.) Let $a \in K^{\times}$ such that $\tilde{a}^{-1}U_0\tilde{a} \subseteq U_0$. We show $a^{-1}Ba \subseteq B$. Let $b \in B$. We distinguish two cases:

a) $|b| = |1|$. Then $\tilde{b} \in U_0$ and so $\tilde{a}^{-1}\tilde{b}\tilde{a} \in \tilde{a}^{-1}U_0\tilde{a} \subseteq U_0$. Thus we obtain $|a^{-1}ba| = |1|$, hence $a^{-1}ba \in B$.

b) $|b| < |1|$. By the principle of domination $|1+b| = |1|$. As in a) we have $a^{-1}(1+b)a = 1 + a^{-1}ba \in B$, hence $a^{-1}ba \in B$.

2.) Now let $a^{-1}Ba \subseteq B$. We show $\tilde{a}^{-1}U_0\tilde{a} \subseteq U_0$. If $\tilde{b} \in U_0$ then b is a unit in B. Then $a^{-1}ba \in a^{-1}Ba \subseteq B$ and $a^{-1}ba$ is a unit in B. Thus $\tilde{a}^{-1}\tilde{b}\tilde{a} \in U_0$.

3.) Hence for $a \in K^{\times}$

$$a^{-1}Ba \subseteq B \iff \tilde{a}^{-1}U_0\tilde{a} \subseteq U_0 .$$

This implies

$$a^{-1}Ba \subseteq b^{-1}Bb \iff \tilde{a}^{-1}U_0\tilde{a} \subseteq \tilde{b}^{-1}U_0\tilde{b}$$

for $a, b \in K^{\times}$. Thus the mapping $a^{-1}Ba \longmapsto \tilde{a}^{-1}U_0\tilde{a}$ is well defined and a bijection.

A valuation ring B of K is said to be **invariant** if $a^{-1}Ba = B$ for all $a \in K^{\times}$.

Clearly, it is sufficient to require $a^{-1}Ba \subseteq B$ for all $a \in K^{\times}$. A valuation is called invariant if its valuation ring is invariant.

PROPOSITION 2.5. The following are equivalent:

(1) B is invariant.

(2) The unit group E of B is invariant.

(3) $U_0 = \{ \text{id} \}$

PROOF : (1) ⇒ (2). Since an inner automorphism maps a unit onto a unit, the invariance of B implies the invariance of E.

(2) ⇒ (3). Since U_0 is the image of E under v and E is invariant, U_0 is a normal subgroup of G. If we consider G as a group acting on W then U_0 is the stabilizer of $|1|$. Since the stabilizer of an element $|a| \in W$, $a \neq 0$, is $\tilde{a}U_0\tilde{a}^{-1}$ we conclude from U_0 being a normal subgroup that U_0 stabilizes every element of W and contains therefore only the identity.

(3) ⇒ (1). $U_0 = \{ \text{id} \}$ has only one conjugate. By 2.4 the same holds for B, hence B is invariant.

THEOREM 2.6. A valuation is invariant if and only if the canonical mapping $\varphi : G \longrightarrow W$ is injective.

PROOF : Since U_0 contains the elements $g \in G$ such that $\varphi(g) = |1|$, the injectivity of φ implies $U_0 = \{ id \}$, hence the valuation is invariant by 2.5.

Conversely, let the valuation be invariant. Then $U_0 = \{ id \}$ by 2.5. When $\varphi(g) = \varphi(h)$ then $\varphi(g^{-1}h) = |1|$ by (B3), hence $g^{-1}h \in U_0$, hence $g = h$. Therefore φ is injective.

When φ is injective we may embed G into W identifying v(a) with $|a|$. Then W becomes a group with null. Hence we obtain a Schilling valuation or in the commutative case a Krull valuation where axiom (B3) is replaced by the multiplicativity of the valuation. Observe that the embedding of the value group into the value set is only possible in the invariant case.

It is sometimes convenient to use exponential valuations instead of valuations. They are obtained by a trivial reformulation of the valuation axioms. Let $| \; | : K \longrightarrow W$ be a valuation. By reversing the order in the value set W and replacing 0 by ∞ and defining $w(a) = |a|$ we obtain a surjective mapping $w : K \longrightarrow W$ satisfying

(B1′) $w(a) = \infty \iff a = 0$.

(B2′) $w(a + b) \geq \mathrm{Min}(w(a), w(b))$.

(B3′) For every $a \in K$ there exists an order preserving mapping $\tilde{a} : W \longrightarrow W$ such that $w(ab) = \tilde{a}w(b)$ for all $b \in K$.

Such a function w satisfying the above axioms is called an **exponential valuation**. The valuation ring associated with an exponential valuation is

$$B = \{ a \in K \mid 0 \leq w(a) \}$$

where w(1) is denoted by 0. Obviously, an exponential valuation leads to a valuation by reversing the order in its value set.

3. The ideal structure of a valuation ring

Let $|\ |: K \longrightarrow W$ be a valuation, B its valuation ring, and G its value group. A subset U of G is called **φ-convex** if $u_1,\ u_2 \in U$, $g \in G$, and

$$\varphi(u_1) \leq \varphi(g) \leq \varphi(u_2)$$

imply $g \in U$. U is called **symmetric** when $id \in U$ and $u \in U$ implies $u^{-1} \in U$.

Let $\mathfrak{F}$ be the set of all φ-convex symmetric subsets of G and $\mathfrak{U}$ the set of all φ-convex subgroups of G. Since every subgroup is symmetric, $\mathfrak{U}$ is a subset of $\mathfrak{F}$.

PROPOSTION 3.1. Let $U \in \mathfrak{F}$. Then

$$I(U) = \{a \in B \mid \ \tilde{a} \notin U\}$$

is a proper two-sided ideal of B. If $U \in \mathfrak{U}$ then I(U) is a completely prime ideal.

PROOF: Let $a \in I = I(U)$, $b \in B$. Then we have

$$|ab| \leq |a| \leq |1|,$$

hence

$$\varphi(\tilde{a}\tilde{b}) \leq \varphi(\tilde{a}) \leq \varphi(id).$$

Suppose $\tilde{a}\tilde{b} \in U$. Since U is φ-convex this implies $\tilde{a} \in U$ contrary to $a \in I(U)$. Therefore, we have $\tilde{a}\tilde{b} \notin U$, hence $ab \in I$.

On the other hand, we have $|1| \leq |a^{-1}| \leq |a^{-1}b^{-1}|$, hence

$$\varphi(id) \leq \varphi(\tilde{a}^{-1}) \leq \varphi(\tilde{a}^{-1}\tilde{b}^{-1}).$$

Using the above argument we obtain $\tilde{a}^{-1}\tilde{b}^{-1} \notin U$, hence $\tilde{b}\tilde{a} \notin U$, hence $ba \in I$. This shows that I is a two-sided ideal. (One should bear in mind that, since B is a valuation ring, we need not show that I is additively closed.)

Since $v(1) = id \in U$, surely $1 \notin I$, hence I is a proper ideal.

Now let $U \in \mathfrak{U}$. Then $a,b \in B$, but a, b not in I, implies $\tilde{a}$, $\tilde{b}$ $\in U$, hence $\tilde{a}\tilde{b} \in U$, hence $ab \notin I$. Therefore, I is completely prime.

THEOREM 3.2. Let J be the set of all proper two-sided ideals of B and $\mathfrak{P}$ the subset of all completely prime ideals. Then the mapping

$$U \longmapsto I(U) = \{ a \in B \mid \tilde{a} \notin U \}$$

is a bijection from $\mathfrak{S}$ onto J. Its restriction on $\mathfrak{U}$ yields a bijection from $\mathfrak{U}$ onto $\mathfrak{P}$.

PROOF: Let $I \in J$. We define

$$U(I) = \{ \tilde{a} \in G \mid a \notin I \text{ and } a^{-1} \notin I \}$$

Obviously, $U = U(I)$ is well-defined and symmetric. It remains to be shown that U is φ-convex. Let $\tilde{a}_1$, $\tilde{a}_2 \in U$, $\tilde{b} \in G$ such that

$$\varphi(\tilde{a}_1) \leq \varphi(\tilde{b}) \leq \varphi(\tilde{a}_2),$$

hence

$$|a_1| \leq |b| \leq |a_2|.$$

It follows $b^{-1}a_1$, $a_2{}^{-1}b \in B$. Since

$$a_1 = b(b^{-1}a_1) \notin I \quad \text{and} \quad a_2{}^{-1} = (a_2{}^{-1}b)b^{-1} \notin I$$

we conclude that neither b nor b^{-1} is in I, hence $\tilde{b} \in U$. Thus $U \in \mathfrak{S}$.

Now let I be a completely prime ideal. $\tilde{a}$, $\tilde{b} \in U$ imply a, b, a^{-1}, $b^{-1} \notin I$ and therefore since I fulfils the condition (P) ab, $(ab)^{-1} \notin I$. It follows $\tilde{a}\tilde{b} \in U$. Hence U is a subgroup of G.

Finally, we have to show that the mapping $I \longmapsto U(I)$ is the inverse function of $U \longmapsto I(U)$. Let $a \in I$. Then $\tilde{a} \notin U(I)$, hence $a \in I(U(I))$. Conversely, let $a \in I(U(I))$. By 3.1 $I(U(I))$ is a proper ideal of B, hence $a \in M$ and therefore $a^{-1} \notin B$, in particular $a^{-1} \notin I$. The assumption $a \notin I$ would imply $\tilde{a} \in U(I)$, hence $a \notin I(U(I))$. Therefore we obtain $a \in I$. Thus, $I = I(U(I))$.

Now let $U \in \mathfrak{S}$. Then $\tilde{a} \in U$ implies $\tilde{a}^{-1} \in U$ since U is symmetric, hence $a \notin I(U)$ and $a^{-1} \notin I(U)$ and therefore $\tilde{a} \in U(I(U))$. On the other hand, $\tilde{a} \notin U$ implies $\tilde{a}^{-1} \notin U$, hence a or a^{-1} belongs to I(U) since a or a^{-1} belongs to B. We conclude $\tilde{a} \notin U(I(U))$. Thus $U = U(I(U))$.

Since the mapping $I \longmapsto U(I)$ is antitone, both $\mathfrak{S}$ and $\mathfrak{U}$ are totally ordered by inclusion. The subgroup U_0 considered above is the least element of $\mathfrak{J}$ and $\mathfrak{U}$ and corresponds to the maximal ideal M of B.

By 1.7 the star-mapping induces a bijection $\mathfrak{I} \longrightarrow \mathbf{O}$ where $\mathbf{O}$ is the set of subrings of K containing B. If we combine this bijection with the bijection of 3.2 we obtain a bijection $\mathfrak{U} \longrightarrow \mathbf{O}$, $U \longmapsto I(U)^* = \{ a \in K^\times \mid \tilde{a} \in U \} \cup B$.

In order to determine the inverse function $\mathbf{O} \longrightarrow \mathfrak{U}$ let $B' \in \mathbf{O}$. As was shown in the proof of 3.2, B' corresponds to

$$U = \{ \tilde{a} \in G \mid a \notin B'^* \text{ and } a^{-1} \notin B'^* \}$$

Taking the definition of the star-mapping into account we have

$$\tilde{a} \in U \iff a \in B' \text{ and } a^{-1} \in B' \iff a \text{ is unit in } B',$$

hence

THEOREM 3.3. The mapping

$$\mathbf{O} \longrightarrow \mathfrak{U},\ B' \longmapsto U = \{ \tilde{a} \in G \mid a \text{ is unit in } B' \}$$

is bijective.

On the other hand, given U we obtain the unit group of B' by $\{ a \in K^\times \mid \tilde{a} \in U \}$. Obviously, the unit group is invariant if and only if U is a normal subgroup of G. Since by 2.5 the invariance of B' is equivalent to the invariance of its unit group, we obtain

COROLLARY 3.4. A valuation ring $B' \in \mathbf{O}$ corresponding to $U \in \mathfrak{U}$ is invariant if and only if U is a normal subgroup of G.

4. Examples of non-invariant valuations

The main purpose of this section is to show the construction of fields with non-invariant valuations. There are two methods we shall use. The first dates back to Hilbert and Hahn, the second to Ore. For more details we refer to Neumann [1] or Cohn [1].

1. Formal power series fields. Let K be a field, G a totally ordered group acting on K as automorphism. We write $a \longmapsto a^g$ for the action of $g \in G$ on the elements $a \in K$.

Let K(G) be the set of all mappings $f: G \longrightarrow K$ whose support

$$T(f) = \{ g \in G \mid f(g) \neq 0 \}$$

is well-ordered with respect to the order defined on G. As usual f is written as a formal series

$$f = \sum_{g \in G} g\, a_g$$

where the coefficients a_g are the images f(g). The sum of two series is defined by adding the coefficients belonging to the same group element g in both series, the product by the formula

$$\sum_{g \in G} g\, a_g \sum_{g \in G} g\, b_g = \sum_{g \in G} g\, c_g$$

with

$$c_g = \sum_{h \in G} a^h_{gh^{-1}}\, b_h$$

One can show that the sum for c_g is actually finite and that K(G) is a field (Neumann [1]).

Since the support T(f), $f \neq 0$, is well-ordered there exists a least element g of T(f). We define g as the order of f denoted by ord f. If f = 0 we put ord f = ∞. The definition of the addition and multiplication of formal series implies

$$\operatorname{ord}(f + g) \geq \operatorname{Min}(\operatorname{ord} f, \operatorname{ord} g),$$

$$\operatorname{ord} fg = \operatorname{ord} f \operatorname{ord} g.$$

Hence ord is an invariant exponential valuation with the value group G. Therefore we obtain (see Schilling [2])

PROPOSITION 4.1. Let G be a totally ordered group. Then there exists a field with an invariant valuation whose value group is G.

PROOF : Let K be an arbitrary field and G act trivially on K. We consider the valuation on K(G) corresponding to the exponential valuation ord.

Remark: 3.4 and 4.1 offer a simple method to construct non-invariant valuations. Let G be a totally ordered group containing a convex subgroup U which is not normal. By 4.1 there exists a field K with an invariant valuation whose value group is G. By 3.4 U corresponds to a non-invariant valuation ring B′. For examples of such groups we refer to Rado [1] or Chehata [1]. The non-invariant valuation rings obtained in this manner are called **subinvariant** and are characterized by the fact that they contain an invariant valuation ring. (Mathiak [5]).

The unit element e of the group G is simultaneously the unit element of the field K(G). By the mapping $a \mapsto ea$ the field K is embedded into K(G). Identifying a with ea, K becomes a subfield of K(G).

We shall show that every valuation $|\ |: K \to W$ may be extended to a valuation $|\ |': K(G) \to W'$. Let $W^\times$ denote the set of non-zero elements of W. We take

$$W'^\times = G \times W^\times$$

ordered by

$$(g_1, w_1) \le (g_2, w_2) :\Leftrightarrow g_2 < g_1 \text{ or } (g_1 = g_2 \text{ and } w_1 \le w_2).$$

Further, in order to get the complete set W′, we add one element (e, 0) as least element to $W'^\times$. W can be embedded, preserving order by mapping w onto (e, w).

THEOREM 4.2. Let $|\ |: K \longrightarrow W$ be a valuation of K, B its valuation ring, and $v: K \longrightarrow \operatorname{End} W$ the homomorphism associated with $|\ |$. Let W' be defined as above. Then

$$|f|' = (k, |a_k|) \quad \text{for } f = \sum_{g \in G} g\, a_g \neq 0,\ k = \operatorname{ord} f$$

$$|f|' = (e, 0) \quad \text{for } f = 0.$$

admits a valuation $|\ |': K(G) \longrightarrow W'$ extending $|\ |$. The as-associated homomorphism: $K(G) \longrightarrow \operatorname{End} W'$ is given by

$$v'(f)(g, w) = (kg, v(a_k{}^g)w)\ ,\ f \neq 0.$$

All the assertions of this theorem are readily established. We leave their verification to the reader.

Slightly easier with respect to the order of the value set is the corresponding theorem for exponential valuations. Then the order of $W'^{\times}$ is just the lexicographic order.

By 4.2 a valuation ring B of K corresponds to a valuation ring

$$B' = \{ f \in K(G) \mid \operatorname{ord} f \geq e \text{ and } a_e \in B \}$$

of $K(G)$. Consider the mapping $B \longmapsto B'$, which is injective since $B = B' \cap K$.

It is possible to determine the set of valuation rings of $K(G)$ which are conjugate to B' only by considering valuation rings in K.

THEOREM 4.3. Every valuation ring which is conjugate to B' can be represented as

$$B_1' = \{ f \in K(G) \mid \operatorname{ord} f \geq e \text{ and } a_e \in B_1 \}$$

where $B_1 = a^{-1}B^g a$, $a \in K^{\times}$, $g \in G$.

PROOF : Let $B'' = f^{-1}B'f$ be a valuation ring of K(G) conjugate to B'. Let $a = a_g$ be the first non-zero coefficient of f. Consider the valuation ring

$$B_1 = a^{-1}B^g a.$$

By a simple computation we obtain

$$f^{-1}hf = k(a^{-1})^k \, b_k{}^g a \quad + \ldots$$

where $h \in K(G)$, $k = \text{ord } h$, and b_k is the first non-zero coefficient of h. Hence ord $f^{-1}hf$ = ord h.

In order to show $f^{-1}B'f \subseteq B_1'$, let $h \in B'$, then $k \geq e$. If $k > e$ then

$$\text{ord } f^{-1}hf = \text{ord } h = k > e,$$

hence $f^{-1}hf \in B_1'$. If $k = e$, then the first coefficient of $f^{-1}hf$ is

$$a^{-1}b_e{}^g a \quad \in B_1.$$

Hence, in any case $f^{-1}hf \in B_1'$. Thus $B'' \subseteq B_1'$.

Similarly, we show that $h \in B_1'$ implies $fhf^{-1} \in B'$, hence $h \in f^{-1}B'f = B''$. Therefore, the inverse inclusion holds, hence $B'' = B_1'$.

In order to simplify the situation we consider a special case. Let K be a commutative field and σ an automorphism of K of finite order. As totally ordered group we consider $G = \mathbb{Z}$. In order to write G multiplicatively we set $g = t^n$ for the elements of G. Instead of K(G) we write K((t)) as usual and the elements as formal power series

$$f = \Sigma \; t^n \, a_n$$

where only a finite number of coefficients with negative subscripts are different from zero. Multiplying two series' one must pay attention to the commutation rule

$$at = t\sigma(a) \qquad \text{for all } a \in K.$$

We assume that a discrete exponential valuation $w: K \longrightarrow \mathbb{Z} \cup \{\infty\}$ is given. Let B be the valuation ring.

Since σ has finite order, only a finite number of distinct valuations

$$w_n = w\,\sigma^n, \quad n = 0, \ldots, r-1.$$

exists. Let B_n be the valuation rings belonging to w_n.

Let W' be the value set of the extended exponential valuation w' on $K((t))$. As mentioned above

$$W'^{\times} = \mathbb{Z} \times \mathbb{Z}$$

is lexicographically ordered.

We shall first determine the complete automorphism group of $\mathbb{Z} \times \mathbb{Z}$, i.e. the group of all bijections $\mathbb{Z} \times \mathbb{Z} \longrightarrow \mathbb{Z} \times \mathbb{Z}$ preserving the lexicographic order of $\mathbb{Z} \times \mathbb{Z}$.

Let F be the set of all functions $f\colon \mathbb{Z} \longrightarrow \mathbb{Z}$. For $f \in F$ and $n \in \mathbb{Z}$ we define $f_n \in F$ by

$$f_n(k) = f(n + k).$$

Let $\mathbb{Z} \times F$ be the set of all pairs (k, f), $k \in \mathbb{Z}$, $f \in F$. Then $\mathbb{Z} \times F$ becomes, as is easily shown, a group with respect to the operation

$$(k, f)\,(n, g) = (k + n,\, f_n + g).$$

PROPOSITION 4.4. The automorphism group of $\mathbb{Z} \times \mathbb{Z}$ is isomorphic to $\mathbb{Z} \times F$.

PROOF : Let u be an automorphism of $\mathbb{Z} \times \mathbb{Z}$ and $H(n)$ the subset of elements of $\mathbb{Z} \times \mathbb{Z}$ whose first component is n. Given two elements of $H(n)$, the number of elements which lie between them is finite. On the other hand, given two elements belonging to distinct subsets $H(n)$, the number of elements which lie between them is infinite. Since u is an automorphism the same is true for the images under u. Hence u maps $H(n)$ onto $H(f(n))$ where $f \in F$.

We claim $f(n + 1) = f(n) + 1$. Since $H(n) \cup H(n + 1)$ is convex, $H(f(n)) \cup H(f(n + 1))$ is also convex. But this is the case if and only if $f(n + 1) = f(n) + 1$.

By induction this implies $f(n + m) = f(n) + m$. Setting $k = f(0)$ we obtain $f(n) = n + k$, hence

$$u(H(n)) = H(n + k)$$

where $k \in \mathbb{Z}$ is independent of n. Therefore

$$u(n, 0) = (n + k, g(n))$$

with $g \in F$. Since the number of elements between $(n, 0)$ and (n, m) is equal to the number of elements between $u(n, 0)$ and $u(n, m)$ we obtain

$$u(n, m) = (n + k, g(n) + m)$$

where $(k, g) \in \mathbb{Z} \times F$ is uniquely determined by u. As is now easily verified, the mapping $u \longmapsto (k, g)$ is an isomorphism of the automorphism group of $\mathbb{Z} \times \mathbb{Z}$ onto $\mathbb{Z} \times F$.

We are now ready to determine the value group as a subgroup of $\mathbb{Z} \times F$. Let

$$f = t^k a_k + \dots, \qquad a_k \neq 0.$$

By 4.2 we have

$$v'(f)(n, m) = (k + n, w(\sigma^n a_k) + m).$$

Comparing this expression with the formula in the proof of 4.4 $v'(f)$ corresponds to the element (k, g) where $g(n) = w(\sigma^n(a_k))$ Therefore g is a r-periodic function $\mathbb{Z} \longrightarrow \mathbb{Z}$.

From the independence of the valuations $w_n = w\,\sigma^n$, $n = 0, \dots, r-1$ follows that every r-periodic function $g \in F$ is obtained. Hence the value group G of w' is isomorphic to the subgroup of $\mathbb{Z} \times F$ containing all elements (n, g) where $g: \mathbb{Z} \longrightarrow \mathbb{Z}$ is r-periodic.

Since every r-periodic function is determined by r values we may identify this subgroup by $\mathbb{Z} \times \mathbb{Z}^r$.

EXAMPLE : Let $K = \mathbb{Q}(i)$, $i^2 = -1$, and σ be the automorphism $a + bi \longmapsto a - bi$. Consider a p-adic exponential valuation, $p \equiv 1 \bmod 4$, on $\mathbb{Q}$ which has two extensions on K denoted by w_0 and w_1. We have $w_1 = w_0\sigma$.

Let B_0 and B_1 be the valuation rings belonging to w_0 and w_1. By 4.3 $\{B_0', B_1'\}$ where

$$B_n' = \{ f = a_0 + t\, a_1 + \dots \mid a_0 \in B_n \},\ n = 0, 1$$

forms the complete system of conjugate valuation rings in K((t)). The corresponding exponential valuations are

$$w_n'(f) = (k, w_n(a_k)), \quad n = 0, 1$$

for $f = t^k a_k + \dots,\ a_k \neq 0$.

The value group is isomorphic to $\mathbb{Z} \times \mathbb{Z}^2$ where the operation on $\mathbb{Z} \times \mathbb{Z}^2$ is defined by

$$(k, g_0, g_1)(n, h_0, h_1) = (k+n, g_0+h_0, g_1+h_1), \quad \text{if } n \text{ is even}$$

$$= (k+n, g_1+h_0, g_0+h_1), \quad \text{if } n \text{ is odd.}$$

The isomorphism of G onto $\mathbb{Z} \times \mathbb{Z}^2$ is given by

$$v'(f) \longmapsto (k, w_0(a_k), w_1(a_k)).$$

If we identify G and $\mathbb{Z} \times \mathbb{Z}^2$ then

$$(k, g_0, g_1) \longmapsto (k, g_0)$$

is the canonical mapping $G \longrightarrow \mathbb{Z} \times \mathbb{Z}$.

REMARK : The center of the field K((t)) is the set of power series

$$f = \Sigma\, t^n a_n, \quad n \equiv 0 \bmod 2, \quad a_n \in \mathbb{Q}.$$

Hence K((t)) has dimension 4 over its center and is therefore a quaternion division algebra.

2. Skew rational function fields. Let W be a totally ordered set with a least element 0. On End W we define a partial order by

$$a \leq b \quad :\Longleftrightarrow \quad aw \leq bw \quad \text{for all } w \in W.$$

More exactly, End W becomes a lattice by this definition.

Namely, the mappings

$$\sup(a, b)w := \mathrm{Max}(aw, bw)$$

$$\inf(a, b)w := \mathrm{Min}(aw, bw)$$

for $a, b \in \mathrm{End}\, W$ are again endomorphisms of W.

Let Aut W denote the automorphism group of W. It is easily shown that Aut W is a lattice ordered group with respect to the order defined above.

The null endomorphism of W maps all elements of W onto the least element 0 and is also denoted by 0. (From the circumstances it will always be clear what is meant.)

Let R be an entire ring. Consider a mapping

$$v: R \longrightarrow \mathrm{Aut}\, W \cup \{0\}$$

satisfying the following conditions

(a) $v(a) = 0 \quad \Leftrightarrow \quad a = 0$

(b) $v(a + b) \leq \sup(v(a), v(b))$

(c) $v(ab) = v(a)v(b)$.

When a valuation $|\ |: K \longrightarrow W$ is given the associated homomorphism v is such a mapping. Conversely, if $v: K \longrightarrow \mathrm{Aut}\, W \cup \{0\}$ with the properties listed above is given and w_0 is a non-zero element of W then by

$$|a| = v(a)w_0$$

a valuation $|\ |: K \longrightarrow W'$ is defined where $W' = \{ v(a)w_0 \mid a \in K \}$. Replacing w_0 by another non-zero element $w_1 \in W'$ and defining $|a|_1 = v(a)w_1$ we obtain a conjugate valuation of K. Namely, let $w_1 = v(b)w_0$ we have

$$\begin{aligned} B_1 &= \{a \in K \mid |a|_1 \leq |1|_1\} \\ &= \{a \in K \mid v(a)v(b)w_0 \leq v(b)w_0\} \\ &= \{a \in K \mid |b^{-1}ab| \leq |1|\} \\ &= bBb^{-1}. \end{aligned}$$

This shows that v represents a complete system of conjugate valuations.

A left **Ore domain** R is an entire ring satisfying the following condition: For $f, g \in R$, $f \neq 0$, there exist $u, v \in R$, $u \neq 0$, such that $uf = vg$. It is well known that every left Ore domain has a left field of fractions

$$K = \{ f^{-1}g \mid f, g \in R, f \neq 0 \}.$$

PROPOSITION 4.5. Let W be a totally ordered set with a least element O. Let R be a left Ore domain and K its left field of fractions. Then a mapping

$$v : R \longrightarrow \text{Aut } W \cup \{O\}$$

satisfying (a), (b), and (c) can be extended to a mapping

$$v' : K \longrightarrow \text{Aut } W \cup \{O\}$$

satisfying (a), (b), and (c).

PROOF : Let $f^{-1}g \in K$. Since $f \neq 0$ we have $v(f) \in \text{Aut } W$, hence $v(f)$ is invertible. We define

$$v'(f^{-1}g) = v(f)^{-1}v(g).$$

We must show that v' is well defined: Let $f^{-1}g = f_1^{-1}g_1$. This means that there exist $u, u_1 \neq 0$ such that

$$uf = u_1f_1, \quad ug = u_1g_1.$$

Putting $\tau = v(f)^{-1}v(g)$ we have

$$v(u_1)v(f_1)\tau = v(u)v(f)\tau = v(u)v(g) = v(u_1)v(g_1),$$

hence $\tau = v(f_1)^{-1}v(g_1)$.

It is easily checked that v' satisfies (a), (b), and (c).

We shall need one further auxiliary result.

PROPOSITION 4.6. Let $\tau: W \longrightarrow W$ be a monomorphism. Then there exists a totally ordered set W^* containing W such that τ is extended to an automorphism $\tau^*: W^* \longrightarrow W^*$.

PROOF : On the set of pairs (w, τ^n), $w \in W$, $n = 0, 1, 2, \ldots$ we define the relation

$$(w_1, \tau^n) \sim (w_2, \tau^k) :\Longleftrightarrow \tau^k w_1 = \tau^n w_2$$

This relation is obviously reflexive and symmetric. In order to show the transitivity assume

$$(w_0, \tau^n) \sim (w_1, \tau^m) \quad \text{and} \quad (w_1, \tau^m) \sim (w_2, \tau^k)$$

hence

$$\tau^m w_0 = \tau^n w_1 \quad \text{and} \quad \tau^k w_1 = \tau^m w_2$$

hence

$$\tau^{m+k} w_0 = \tau^{n+k} w_1 = \tau^{m+n} w_2 .$$

Since τ is injective we obtain $\tau^k w_0 = \tau^n w_2$ and therefore $(w_0, \tau^n) \sim (w_2, \tau^k)$.

Hence, the relation defined above is an equivalence relation. Let W^* be the set of equivalence classes and denote by w/τ^n the equivalence class representing (w, τ^n).

We define

$$w/\tau^n \leq w'/\tau^k \quad :\Longleftrightarrow \quad \tau^k w \leq \tau^n w' .$$

Thereby W^* becomes a totally ordered set with the least element $0/\tau^0$. The mapping $w \longmapsto w/\tau^0$ is a monomorphism from W into W^*. Identifying $w \in W$ with w/τ^0 we get W as a subset of W^*.

Now it is easy to show that the mapping

$$\tau^*: W^* \longrightarrow W^*, \ \tau^*(w/\tau^n) = (\tau w)/\tau^n$$

is an automorphism of W^* extending τ.

Let $\sigma: K \longrightarrow K$ be a monomorphism of the field K. The set of polynomials

$$K[x, \sigma] = \{ a_0 + a_1 x + \ldots + a_n x^n \mid a_i \in K \}$$

is a ring when two polynomials are multiplied by using the commutation rule

$$\sigma(a)x = xa .$$

$K[x, \sigma]$ is called **skew polynomial ring** over K. It reduces to the ordinary polynomial ring with central indeterminate when σ is the identity. It is well known that $K[x, \sigma]$ is a left Ore domain (Ore [1]). Hence $K[x, \sigma]$ has a left field of fractions which is denoted by $K(x, \sigma)$. We call $K(x, \sigma)$ a **skew rational function field** or shorter an **Ore field** over K.

A monomorphism $\sigma: K \longrightarrow K$ is called **compatible** with a valuation $| \ |: K \longrightarrow W$ when there exists a monomorphism $\tau: W \longrightarrow W$ such that $|\sigma a| = \tau|a|$ for all $a \in K$.

THEOREM 4.7. A valuation $K \longrightarrow W$ can be extended to a valuation $K(x, \sigma) \longrightarrow W^*$ if σ is compatible with the valuation of K. For polynomials the formula

$$|a_0 + a_1x + \ldots + a_nx^n| = \mathrm{Max}(|a_k|)$$

is valid.

PROOF : 1.) Let $v: K \longrightarrow \mathrm{Aut}\, W \cup \{0\}$ be the homomorphism associated with the valuation of K. Let W^* be the extension of W constructed in 4.6.

First we show that $v(a): W \longrightarrow W$ can be extended to an order preserving mapping $v_0(a): W^* \longrightarrow W^*$. We define

$$v_0(a)(w/\tau^n) = v(\sigma^n a)w/\tau^n.$$

In order to show that $v_0(a)$ is well defined assume

$$|b|/\tau^n = |c|/\tau^k, \quad b, c \in K.$$

Then we have $\tau^k|b| = \tau^n|c|$, hence $|\sigma^k b| = |\sigma^n c|$ because σ is compatible with the valuation. Then a short calculation yields

$$\tau^k(v(\sigma^n a)|b|) = \tau^n(v(\sigma^k a)|c|),$$

hence

$$v(\sigma^n a)|b|/\tau^n = v(\sigma^k a)|c|/\tau^k.$$

Immediately from the definition follows

$$v_0(ab) = v_0(a)v_0(b) \quad \text{and } v_0(1) = \mathrm{id}.$$

Therefore, $v_0(a)$ is bijective for $a \in K^{\times}$. Further, $v_0(a)$ preserves the order of W^*, hence $v_0(a) \in \mathrm{Aut}\, W^*$ for $a \in K^{\times}$. Now it is easily shown that

$$v_0 : K \longrightarrow \mathrm{Aut}\, W^* \cup \{0\}$$

satisfies the conditions (a), (b), (c).

2.) By 4.6 there exists an automorphism $\tau^* : W^* \longrightarrow W^*$ extending τ. Let $f(x) = a_0 + a_1 x + \ldots + a_n x^n \in K[x, \sigma]$. We define

$$v_1(f(x)) = \sup_k v_0(a_k)\tau^{*k}.$$

Obviously, v_1 is an extension of v_0. One checks that v_1 is a mapping

$$v_1 : K[x, \sigma] \longrightarrow \mathrm{Aut}\, W^* \cup \{0\}$$

satisfying (a), (b), and (c). (In the proof of (c) one needs the formula

$$v_0(\sigma a)\tau^* = \tau^* v_0(a) \quad \text{for all } a \in K$$

which is obtained by a simple calculation.)

3.) By 4.5

$$v_2 : K(x, \sigma) \longrightarrow \mathrm{Aut}\, W^* \cup \{0\}, \qquad v_2(f^{-1}g) = v_1(f)^{-1} v_1(g)$$

is an extension onto the Ore field satisfying (a), (b), (c).

By

$$|f^{-1}g| := v_2(f^{-1}g)|1|$$

we obtain an extension of the valuation of K onto $K(x, \sigma)$. The valuation axioms follow from (a), (b), (c). Finally we must show that any element $|a|/\tau^n \in W^*$ is a value of an element of $K(x, \sigma)$. Let $f(x) = x^n$ and $g(x) = a$. Then

$$|f^{-1}g| = v_1(x^n)^{-1} v_1(a)|1| = |a|/\tau^n.$$

In the special case of a polynomial $f(x) = a_0 + \ldots + a_n x^n$ we have

$$|f| = v_1(f)|1| = \sup_k v_1(a_k)\tau^{*k}|1| = \mathrm{Max}|a_k|.$$

Example: Let k be a commutative field and $K = k(t)$ a rational function field over k (with central indeterminate t). Consider the t-adic exponential valuation on K

$$w: K \longrightarrow W = \mathbb{Z} \cup \{\infty\}$$

i.e. if $a(t) = t^k a^*(t)$, $k \in \mathbb{Z}$, with

$$a^*(t) = \frac{a_0 + a_1 t + \dots + a_n t^n}{b_0 + b_1 t + \dots + b_m t^m}, \quad a_0, b_0 \neq 0$$

then $w(a(t)) = k$ and $w(0) = \infty$.

The homomorphism associated with w is given by

$$v(a(t))u = u + w(a(t)), \quad u \in W.$$

Let $K(x, \sigma)$ be the Ore field over K where σ is the monomorphism

$$\sigma: K \longrightarrow K, \ a(t) \longmapsto a(t^2).$$

Defining $\tau: W \longrightarrow W$ by $\tau w = 2w$ we obtain

$$w(\sigma(a(t))) = w(a(t^2)) = \tau w(a(t)),$$

hence, σ is compatible with the t-adic valuation.

In a first step we must enlarge W to W^* such that τ is extended to an automorphism of $W^* = \{ w/\tau^n \mid w \in W, n = 0, 1, \dots \}$. It is easily shown that W^* can be identified with $\mathbb{Z}[1/2] \cup \{\infty\}$ where $\mathbb{Z}[1/2]$ is the set of all rational numbers whose denominators are powers of 2. The extension of τ is $\tau^*: W^* \longrightarrow W^*$, $\tau^* u = 2u$. The order on W^* is the usual order of rational numbers.

By 4.7 there exists an extension $w^*: K(x, \sigma) \longrightarrow W^*$ of $w: K \longrightarrow W$.

First we describe w^* explicitely for a polynomial

$$h(x) = a_0(t) + a_1(t)x + \dots + a_n(t)x^n$$

By the formula in 4.7 we have

$$v_1(h(x))u = \inf_i (v(a_i(t))\tau^{*i})u$$

$$= \min_i (2^i u + w(a_i(t))).$$

(Observe that by the use of exponential valuation sup is replaced by inf.)

Let us consider for instance the polynomial

$$h(x) = t^3 + t^{-1}x + t^4x^3.$$

Then we have

$$v_1(h(x))u = \min(u+3,\ 2u-1,\ 8u+4).$$

The graph of $v_1(h(x))$ is monotone increasing and piecewise linear. More exactly, W^* can be divided into a finite set of subintervals on each of which $f = v_1(h(x))$ is a linear function

$$f(u) = 2^k u + n, \quad k \in \mathbf{Z}, \quad n \in \mathbf{Z}[1/2]$$

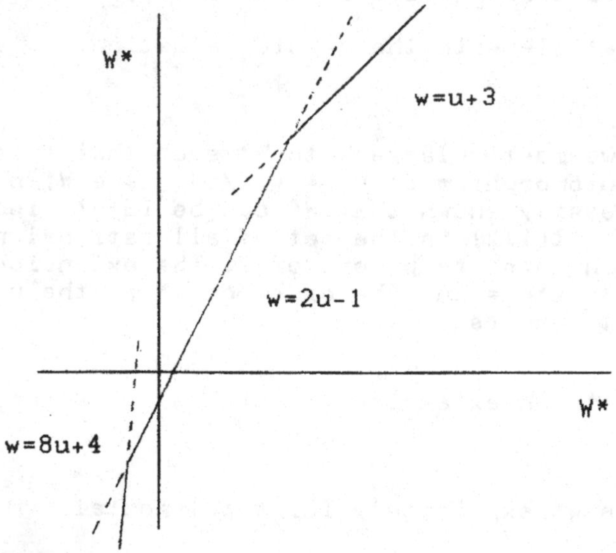

Let S be a the set of the rational numbers 2^n, $n \in \mathbb{Z}$. A function $f:W^* \longrightarrow W^*$ is called piecewise S-linear if there exist elements $a_1, \ldots, a_k \in W^*$ such that

$$a_0 = -\infty < a_1 \leq \ldots \leq a_k < a_{k+1} = \infty$$

and $m_0, \ldots, m_k \in S$ and $n_0, \ldots, n_k \in W^*$ such that

$$f(u) = m_i u + n_i \quad \text{for} \quad a_i \leq u \leq a_{i+1}$$

$i = 0, \ldots, k$.

Since $m_i \in S$ are positive, f is monotone increasing and bijective. It is easy to see that the set of all functions of this kind forms a group where the group operation is the composition of functions. We denote this group by G and we will show that G is the value group of the valuation.

PROPOSITION 4.8. The set G_1 of all functions $f \in G$ with only one break is a system of generators of G.

PROOF : This is shown by induction on the numbers of breaks of $f \in G$. Let f be described by a_i, m_i, n_i, $i = 0, \ldots, k$, as above. Consider $g \in G_1$ defined by

$$g(u) = m_0 u + n_0 \quad \text{if} \quad u \leq a_1$$

$$= m_1 u + n_1 \quad \text{if} \quad u \geq a_1.$$

Then fg^{-1} has breaks in $g(a_2), \ldots, g(a_k)$ and is therefore an element of G with one break less than f. Hence we can apply the induction hypothesis.

A function $f \in G_1$ can be represented by

$$f(u) = 2^{-k}(2^{k_0}u + n_0) \quad \text{if} \quad u \leq a$$

$$= 2^{-k}(2^{k_1}u + n_1) \quad \text{if} \quad u \geq a.$$

where k, k_0, k_1 are integers ≥ 0 and n_0, n_1 are arbitrary integers.

f is said to be convex when $k_0 \geq k_1$. If f is not convex, then f^{-1} is convex. Hence G is generated by the convex functions $f \in G_1$. If f is convex, we have

$$f = v_2(g^{-1}h)$$

with

$$h(x) = t^{n_0}x^{k_0} + t^{n_1}x^{k_1} \quad \text{and} \quad g(x) = t^k.$$

On the other hand, we have $v_1(h) \in G$ for a polynomial $h(x) \in K[x, \sigma]$, hence

$$v_2(g^{-1}h) = v_1(g)^{-1}v_1(h) \in G$$

for an arbitrary element $g^{-1}h \in K(x, \sigma)$.

This shows that the value group of the valuation is G. The canonical mapping φ: $G \longrightarrow W^*$ is given by $f \longmapsto f(0)$.

Applying 3.2 we can determine the ideal lattice of the associated valuation ring B. It is easily checked that the set $\mathfrak{F}$ of all symmetric φ-convex subsets of G contains only two members

$$U_0 = \{ f \in G \mid f(0) = 0 \} \quad \text{and} \quad G.$$

By 3.2 there exists only one non-trivial two-sided ideal, namely the maximal ideal. Further, this is the only completely prime ideal of B beside the null ideal. Hence B is a maximal non-invariant valuation ring of K.

EXERCISES

1.) Let B be a valuation ring of a field K and M its maximal ideal. Let $a \in K^\times$. Show $(aB)^* = Ma^{-1}$.

2.) A non-empty subset H of the value set W of a valuation is called a lower class if $w_1 \leq w_2$ and $w_2 \in H$ imply $w_1 \in H$. Prove

a) The mapping $H \longmapsto I = \{ a \in K \mid |a| \in H \}$ is a bijection of the set of all lower classes onto the chain of all fractional right ideals.

b) The mapping $H \longmapsto I = \{ a \in K \mid v(a)h \leq |1| \text{ for all } h \in H\}$ is a bijection of the set of all lower classes onto the chain of all fractional left ideals.

3.) Let B be a valuation ring of K and M its maximal ideal. Prove that the following statements are equivalent.

(1) B is invariant.

(2) M is invariant.

(3) Every right and every left ideal is two-sided.

4.) Let $|\ |$ be a valuation on K. Prove that the following statements are equivalent.

(1) $|\ |$ is invariant.

(2) $|a| = |b|$ implies $|a^{-1}| = |b^{-1}|$ for all $a, b \in K^\times$.

(3) $|a| = |b|$ implies $|ac| = |bc|$ for all $a, b, c \in K$.

(4) $|ab| = |1|$ implies $|ba| = |1|$ for all $a, b \in K^\times$.

5.) Let $|\ |$ be a valuation on K, B its valuation ring, and G its value group. Prove that the following statements are equivalent.

(1) B contains an invariant valuation ring of K.

(2) The set of all valuation rings of K being conjugate to B is totally ordered by inclusion.

(3) G is totally ordered.

CHAPTER 2 VALUATION TOPOLOGIES AND THE APPROXIMATION THEOREM

First we give a short account of V-topologies. In the non-commutative case a valuation topology need not be a V-topology. A valuation inducing a V-topology is called a V-valuation. We shall list a number of conditions under which a valuation induces a V-topology.

The discussion of the approximation theorem begins with a theorem reducing the statement from n valuations to two valuations. Its proof is based on a lemma which is essentially due to Bourbaki [1].

For the approximation theorem additional assumptions on the valuations are necessary. We have to distinguish different notions of independence which agree with each other in the invariant case.

5. V-topologies

Consider following conditions on a filter τ of a field K.

(A) The intersection of all members of τ contains only 0.

(B) For every $U \in \tau$ there exists $V \in \tau$ such that $V - V \subseteq U$.

(C) For every $U \in \tau$ there exists $V \in \tau$ such that $V V \subseteq U$.

(D) (1) For every $U \in \tau$ and $a \in K^{\times}$ there exists $V \in \tau$ such that $aV \subseteq U$.

(2) For every $U \in \tau$ and $a \in K^{\times}$ there exists $V \in \tau$ such that $Va \subseteq U$.

(E) For every $U \in \tau$ there exist $V \in \tau$ such that $(1 + V)^{-1} \subseteq 1 + U$.

Let τ be a filter on K satisfying (A) and (B). Call a subset A of K open if for every $a \in A$ there exists $U \in \tau$ such that $a + U \subseteq A$. By this definition τ induces a Hausdorff topology on K such that K becomes a topological group with respect to addition. A filter τ satisfying (A) - (D) induces a ring topology on K. If τ further satisfies (E) then τ induces a field topology (see van der Waerden [1]).

A filter τ satisfying (A) - (C) is called a **V-filter** if τ satisfies the condition

(V) For every $U \in \tau$ there exists $V \in \tau$ such that $ab \in V$ implies $a \in U$ or $b \in U$.

A topology on K induced by a V-filter is said to be a **V-topology**.

Note that if a filter base of τ satisfies any of the conditions (A) - (D) or (V) then τ itself satisfies this condition.

A subset A of K is called **left bounded** with respect to a filter τ if for every $U \in \tau$ there exists $V \in \tau$ such that $VA \subseteq U$. Analogously, a right bounded set is defined. A subset is called bounded if it is left and right bounded. It is easy to show that a subset of a left bounded set is left bounded and that the sum $A + B$ and the product $A\,B$ of two left bounded sets A and B are left bounded.

A topology induced by a filter τ is called **locally bounded** if τ contains a bounded set.

A subset A is said to be **bounded away from 0** if there exists a set $U \in \tau$ which is disjoint from A.

PROPOSITION 5.1. Let τ be a filter satisfying (A) - (C). Then the following are equivalent.

(1) τ satisfies the condition (V).

(2) Let A be any subset being bounded away from 0. Then

$$A^{-1} = \{ a \in K^{\times} \mid a^{-1} \in A \}$$

is left bounded with respect to τ.

PROOF : (1) $\Rightarrow$ (2). Let $U \in \tau$ and A be a subset of K which is disjoint from $U_1 \in \tau$. We may assume $U_1 \subseteq U$. Since τ is a V-filter there exists $V \in \tau$ such that $ab \in V$ implies $a \in U_1$ or $b \in U_1$. Let $a \in A^{-1}$, hence $a^{-1} \in A$. Since $A \cap U_1 = \emptyset$ we have $a^{-1} \notin U_1$. Therefore $v = (va)a^{-1} \in V$ implies $va \in U_1 \subseteq U$. Thus $V\,A^{-1} \subseteq U$ and A^{-1} is left bounded.

(2) $\Rightarrow$ (1). Let $U \in \tau$. Then the complement $\complement U$ is bounded away from 0, hence $(\complement U)^{-1}$ is left bounded by (2). Hence there exists $V \in \tau$ such that $V(\complement U)^{-1} \subseteq U$. Let $ab \in V$ and $b \notin U$ then $a = (ab)b^{-1} \in V(\complement U)^{-1} \subseteq U$. Thus τ is a V-filter.

REMARK : The notion of a V-topology was introduced by Kaplansky using condition (2) of 5.1. Since the condition (1) is symmetric with respect to left and right multiplication we may replace left bounded by right bounded or even by bounded in condition (2).

PROPOSITION 5.2. A V-topology is a locally bounded field topology.

PROOF : Let τ be a V-filter on K. We prove

(D): Let $a \in K^{x}$ and $U \in \tau$. By (A) there exists $U_1 \in \tau$ such that $a^{-1} \notin U_1$. We may assume $U_1 \subseteq U$, otherwise consider $U_1 \cap U$ instead of U_1. By (V) there exists $V \in \tau$ such that $ab \in V$ implies $a \in U_1$ or $b \in U_1$. Therefore, from $v = a^{-1}(av) = (va)a^{-1} \in V$ we conclude $av, va \in U_1 \subseteq U$, hence $aV, Va \subseteq U$.

(E): Let $u \in \tau$. By (A) and (B) there exist U_1 and U_2 in τ such that

$$(1 + U_1) \cap -U_2 = \emptyset \quad \text{and} \quad U_2 \subseteq U.$$

By (V) there exists $V_1 \in \tau$ such that $ab \in V_1$ implies $a \in U_2$ or $b \in U_2$. Let $V = V_1 \cap U_1$ and $v \in V$. Since $1 + v \in 1 + U_1$ we have $-(1 + v) \notin U_2$. Thus

$$v = ((1 + v)^{-1} - 1)(-(1 + v)) \in V \subseteq V_1$$

implies $(1 + v)^{-1} - 1 \in U_2 \subseteq U$, hence $(1 + V)^{-1} \subseteq 1 + U$.

The topology is locally bounded: Let U be an element of τ such that $1 \notin U$. By (C) there exists $V \in \tau$ such that $V V \subseteq U$ Let $v \in V$, $v \neq 0$. Then $v^{-1} \in \complement V$ since $v^{-1} \in V$ would imply $1 = vv^{-1} \in U$. Therefore V is contained in the bounded set $(\complement V)^{-1} \cup \{ 0 \}$. Hence V is bounded.

Under set-theoretic inclusion the filters on K form a partially ordered set.Then we have

LEMMA 5.3. Let σ, τ be filters on K satisfying (A) - (D). If there exists $V \in \sigma$ which is left bounded with respect to τ, then $\tau \subseteq \sigma$.

PROOF : Let $U \in \tau$. Since V is left bounded there exists $U' \in \tau$ such that $U' V \subseteq U$. Let $a \in U'$, $a \neq 0$. By (D) there exists $V' \in \sigma$ such that $a^{-1}V' \subseteq V$. Then from $V' \subseteq aV \subseteq U'V \subseteq U$ follows $U \in \sigma$.

THEOREM 5.4. V-topologies are minimal among the ring topologies on K.

PROOF : Let τ be a V-filter, σ a filter which generates a ring topology, and $\sigma \subseteq \tau$. By (A) and (C) there exists $U \in \sigma$ such that $1 \notin U\,U$, hence $U \cap U^{-1} = \emptyset$. Since $\sigma \subseteq \tau$ we have $U \in \tau$. By 5.2 U is left bounded with respect to τ, hence $\tau \subseteq \sigma$ by lemma 5.3 and therefore $\sigma = \tau$.

COROLLARY 5.5. Let σ, τ be two distinct V-filters. Then we have for all $U \in \tau$ and $V \in \sigma$

$$U \cap (1 + V) \neq \emptyset$$

PROOF : Assume $U \cap (1 + V) = \emptyset$ for some $U \in \tau$ and $V \in \sigma$. Since by 5.1 σ induces a field topology, there exists $V' \in \sigma$ such that $(1 + V')^{-1} \subseteq 1 + V$. Hence $(1 + V')^{-1}$ is bounded away from 0 and therefore $1 + V'$ is left bounded with respect to τ. By lemma 5.3 we conclude $\tau \subseteq \sigma$. By 5.4 σ is minimal, hence $\tau = \sigma$. We obtain a contradiction.

6. Topologies induced by valuations

Let $| \; |: K \longrightarrow W$ be a non-trivial valuation. We define a topology on K by using the ε-neighbourhoods of 0

$$U(\varepsilon) = \{c \in K \mid |c| \leq \varepsilon \}, \quad \varepsilon \in W^{x}.$$

First we shall show that $U(\varepsilon)$ is a fractional right principal ideal. Choose $a \in K^{x}$ such that $\varepsilon = |a|$. (One should bear in mind that a valuation is a surjective mapping.) For $u \in U(\varepsilon)$ we have $|u| \leq \varepsilon = |a|$, hence $a^{-1}u \in B$, hence $u \in aB$. Conversely, let $u = ab \in aB$. Then we have $|u| = \tilde{a}|b| \leq \tilde{a}|1| = |a| = \varepsilon$, hence $u \in U(\varepsilon)$. Thus

$$U(\varepsilon) = aB, \text{ where } \varepsilon = |a|.$$

Let τ be the filter generated by the filter base

$$\{ aB \mid a \in B, a \neq 0 \}$$

We shall show that τ satisfies the conditions (A) - (C) and (D1). Thus (K,+) becomes a topological group.

Since the valuation is non-trivial there exists $a \in K^{x}$ such that $|a| \neq |1|$. Then the set of values $|a^n|$, $n \in \mathbb{Z}$, and therefore W^{x} has no least element. Hence the intersection of all ε-neighbourhoods is null. Since

$$aB - aB \subseteq aB \quad \text{and} \quad aB\, aB \subseteq aB$$

hold for $a \in B$ and for given $a, b \in K^{x}$ there exists $ba^{-1} \in K^{x}$ such that

$$(ba^{-1})\, aB \subseteq bB,$$

τ satisfies (B), (C), and (D1). It should be mentioned that in general (D2) and (E) are not valid.

Since the non-zero right principal ideals aB, $a \in B$, $a \neq 0$, form a filter base of τ we shall say that the valuation ring B generates the topology and that two valuation rings (resp. the associated valuations) are **topologically equivalent** when they generate the same topology. The definition of the valuation topology by right ideals shows again the asymmetry of a valuation with respect to multiplication.

PROPOSITION 6.1. Two valuation rings B_1 and B_2 of K are topologically equivalent if and only if there exist u, v $\in K^\times$ such that

$$uB_1 \subseteq B_2 \quad \text{and} \quad vB_2 \subseteq B_1$$

PROOF : Every aB_2 contains $(au)B_1$, every aB_1 contains $(av)B_2$, hence $\{ aB_1 \mid a \in K^\times \}$ and $\{ aB_2 \mid a \in K^\times \}$ generate the same filter τ. Thus the induced topologies are equal.

Conversely, the valuation rings B_1 and B_2 are neighbourhoods of 0 since they generate the same topology. Hence there exist u, v $\in K^\times$ such that $uB_1 \subseteq B_2$ and $vB_2 \subseteq B_1$.

COROLLARY 6.2. Let B_1 and B_2 be valuation rings contained in a valuation ring $B \neq K$. Then B_1 and B_2 are topologically equivalent.

PROOF : Let M be the maximal ideal of B. Since M is a completely prime ideal of B_1 and B_2 we have $uB_1 \subseteq B_2$ and $uB_2 \subseteq B_1$ for any $u \in M$, $u \neq 0$.

THEOREM 6.3. Let B be a valuation ring of K. Then the following are equivalent.

(1) B generates a V-topology on K.

(2) B generates a field topology on K.

(3) B generates a ring topology on K.

(4) The valuation rings $a^{-1}Ba$, $a \in K^\times$, generate the same topology as B.

(5) Every non-zero left ideal contains a non-zero right ideal.

(5′) Every non-zero right ideal contains a non-zero left ideal.

PROOF : (1) ⇒ (2) follows by 5.1, (2) ⇒ (3) is trivial.

(3) ⇒ (4). Let $a \in K^\times$. By (3) the mappings $x \mapsto xa$ and $x \mapsto xa^{-1}$ are continuous. Hence there exist u, v $\in K^\times$ such that

$$vBa \subseteq aB \quad \text{and} \quad uBa^{-1} \subseteq a^{-1}B.$$

It follows

$$(a^{-1}va)a^{-1}Ba \subseteq B \quad \text{and} \quad uB \subseteq a^{-1}Ba.$$

Hence $a^{-1}Ba$ generates the same topology as B by 6.1.

(4) ⇒ (5). Let I be a non-zero left ideal and $a \in I$, $a \neq 0$. Since B and $a^{-1}Ba$ are topologically equivalent there exists $u \in K^{\times}$ such that $uB \subseteq a^{-1}Ba$, hence $auB \subseteq Ba \subseteq I$. Therefore we get a non-zero right ideal auB contained in I.

(5) ⇒ (5'). Let I be a non-zero and $a \in I$, $a \neq 0$. Let M be the maximal ideal of B. Then Ma is a non-zero left ideal which by (5) contains a non-zero right ideal J. Let $b \in J$, $b \neq 0$. We obtain $bM \subseteq Ma$ and by applying the star-mapping $a^{-1}B \subseteq Bb^{-1}$, hence $Bb \subseteq aB \subseteq I$. Thus Bb is a non-zero left ideal which is contained in I.

(5') ⇒ (5) follows similarly.

(5) ⇒ (1). For the complement of aB, $a \in K^{\times}$ we obtain the formula

$$(\complement aB)^{-1} = \{ c \in K^{\times} \mid c^{-1} \notin aB \}$$

$$= \{ c \in K^{\times} \mid ca \in M \} = Ma^{-1} \setminus \{0\}.$$

Let A be a subset of K disjoint from aB. Then $A \subseteq \complement aB$, hence $A^{-1} \subseteq (\complement aB)^{-1} \subseteq Ma^{-1}$. By (5) there exists $b \neq 0$ such that $bB \subseteq Ba$. Let U be any neighbourhood of 0 and $uB \subseteq U$, $u \neq 0$. Then we have

$$(ubB)A^{-1} \subseteq (ubB)Ma^{-1} \subseteq uB \subseteq U.$$

This shows that A^{-1} is left bounded. By 5.2 B generates a V-topology.

A valuation is called a **V-valuation** when its valuation ring satisfies the equivalent conditions of 6.3. The second example of section 4 is a valuation which is not a V-valuation. Invariant valuations are V-valuations by 6.3 (4).

If a non-zero right ideal I contains a left ideal $J \neq 0$, then it also contains the two-sided ideal $BJ \neq 0$. Therefore we obtain

COROLLARY 6.4. A topology induced by a V-valuation possesses a filter base of neighbourhoods of 0 consisting of non-zero two-sided ideals.

The following theorem gives a purely valuation theoretical characterisation of V-valuations.

THEOREM 6.5. Let $|\ |: K \longrightarrow W$ be a valuation. The following assertions are equivalent.

(1) $|\ |$ is a V-valuation.

(2) For every $\varepsilon \in W^{x}$ there exists $\varepsilon' \in W^{x}$ such that $|a| \leq \varepsilon'$ implies $|a^{-1}| > \varepsilon$ for all $a \in K^{x}$.

PROOF : (1) ⇒ (2). Let $\varepsilon \in W^{x}$ and $u \in K^{x}$ such that $|u| = \varepsilon$. Since $|\ |$ is a V-valuation, 6.3 implies that the fractional left ideal Mu^{-1} contains a right ideal vB, $v \in K^{x}$. Put $\varepsilon' = |v|$. If $a \in K^{x}$ and $|a| \leq \varepsilon'$ then $a \in vB$, hence $a \in Mu^{-1}$. By the formula in the proof of 6.3 it follows that $a \in (CuB)^{-1}$, hence $a^{-1} \notin uB$, hence $|a^{-1}| > |u| = \varepsilon$.

(2) ⇒ (1). Let I be a non-zero left ideal and b a non-zero element of I. Put $\varepsilon = |b^{-1}|$. By (2) there exists $\varepsilon' \in W^{x}$ such that $|a| \leq \varepsilon'$ implies $|a^{-1}| > \varepsilon$. The right ideal $I' = \{a \in K \mid |a| \leq \varepsilon'\}$ is contained in I, for $|a| \leq \varepsilon'$ implies $|a^{-1}| > \varepsilon = |b^{-1}|$, hence $a^{-1} \notin b^{-1}B$. By the formula we obtain

$$a \in (Cb^{-1}B)^{-1} \subseteq Mb \subseteq I.$$

Thus $I' \subseteq I$ and condition (5) of 6.3 is satisfied. Hence $|\ |$ is a V-valuation.

7. The intersection of a finite number of valuation rings

Let $| \ |_1, \dots, | \ |_n$ be valuations of K, $B_1, \dots, B_n$ their valuation rings and let

$$R = B_1 \cap \dots \cap B_n$$

be the intersection of the valuation rings. In the following we shall investigate the ideal structure of the ring R. (Conf. Krull [1].)

LEMMA 7.1. Let $a \in K^x$. Then there exists an element $b \neq 0$ in R such that $ab = ba \in R$ and

$|b|_k = |a|_k$ for those indices k for which $|a|_k \leq |1|_k$

$|b|_k \leq |a^{-1}|_k$ for those indices k for which $|a|_k > |1|_k$.

PROOF : Let P be the set of all monic polynomials

$$p(t) = t^k + t^{k-1}c_{k-1} + \dots + t\,c_1 + 1$$

where the coefficients c_i are integers. Let I denote the set of indices for which there exists a polynomial $p \in P$ such that $|p(a)|_k < |1|_k$. For each $k \in I$ choose such a polynomial p_k and put

$$p(t) = 1 + t^2 \prod_{k \in I} p_k(t).$$

Obviously, $p \in P$ and $p(a) \neq 0$. We claim that $b = a(p(a))^{-1}$ satisfies the conditions of the lemma.

1.) Let i be an index for which $|a|_i \leq |1|_i$. Then $a \in B_i$ and therefore $p(a) \in B_i$. We first consider the case where $i \in I$. Then $p_i(a) \in M_i$, hence $p(a)$ is a unit in B_i. When $i \notin I$ then there exists no polynomial $p \in P$ such that $p(a) \in M_i$. Hence $p(a)$ is a unit in B_i in any case. We have therefore

$$|b|_i = |a(p(a))^{-1}|_i = |a|_i .$$

2.) Let i be an index such that $|a|_i > |1|_i$, hence $a^{-1} \in M_i$. Since

$$p(t) = t^k + t^{k-1} c_{k-1} + \dots + 1$$

is a polynomial of degree $k \geq 2$ we have

$$b = a(p(a))^{-1} = a^{1-k}(1 + a^{-1}c_{k-1} + \dots + a^{-k})^{-1}$$

hence

$$|b|_k = |a^{1-k}|_k \leq |a^{-1}|_k .$$

For all k we have $|b|_k$, $|ab|_k \leq |1|_k$, hence b, ab $\in$ R. The definition of b shows immediately ab = ba.

COROLLARY 7.2. K is the left and right quotient field of R.

PROOF : Let $a \in K^x$. Defining b as in the lemma we have $a = (ab)b^{-1} = b^{-1}(ba)$ where ab = ba, $b \in R$.

COROLLARY 7.3. Let B_1 not be contained in B_k for $k > 1$. Then there exists $a \in K$ such that

$$|a|_1 = |1|_1, \quad |a|_2 < |1|_2, \quad \dots , \quad |a|_n < |1|_n .$$

PROOF : Since B_1 is not contained in B_k there exists $a_k \in B_1$ such that $|a_k|_k > |1|_k$ for $k = 2, \dots , n$. We may assume $|a_k|_1 = |1|_1$, otherwise replace a_k by $1 + a_k$. By 7.1 there exist $b_2, \dots , b_n \in R$ such that $|b_k|_1 = |1|_1$ and $|b_k|_k \leq |a_k^{-1}|_k < |1|_k$. Then $a = b_2 \dots b_n$ satisfies the conditions of the corollary.

Now we recall some simple facts about localisation. Let R be a subring of a field K.

A non-empty subset S of R is called a **denominator set** in R if

(1) S is multiplicatively closed and $0 \notin S$ and

(2) S satisfies the Ore-condition: For $a \in R$, $s \in S$ there exist $b \in R$, $t \in S$ such that at = sb.

If S is a denominator set in R, then

$$RS^{-1} = \{ as^{-1} \mid a \in R, s \in S \}$$

is a subring of K containing R. When I is a right ideal of R then

$$IS^{-1} = \{ as^{-1} \mid a \in I, s \in S \}$$

is a right ideal of RS^{-1}. Lattice operations are preserved, i.e.

$$(I + J)S^{-1} = IS^{-1} + JS^{-1}$$

$$(I \cap J)S^{-1} = IS^{-1} \cap JS^{-1}$$

where I and J are right ideals of R.

Now let R be the intersection of the valuation rings B_1, ... , B_n. Then we have

PROPOSITION 7.4. The sets

$$S_k = \{ s \in R \mid |s|_k = |1|_k \}$$

are denominator sets in R such that

$$RS_k^{-1} = B_k$$

PROOF : 1.) The property (1) of a denominator set follows immediately from the definition of S_k. In order to prove (2) let $a \in R$, $a \neq 0$, and $s \in S_k$. By the lemma there exists $b \in R$ such that

$$|b|_i = |s^{-1}a|_i \quad \text{if} \quad |s^{-1}a|_i \leq |1|_i$$

$$|b|_i \leq |a^{-1}s|_i \quad \text{if} \quad |s^{-1}a|_i > |1|_i .$$

Put $t = a^{-1}sb$. Then $|s^{-1}a|_i \leq |1|_i$ implies $|t|_i = |a^{-1}sb|_i = |1|_i$. This holds in particular for $i = k$, since $|a|_k \leq |1|_k = |s|_k$ implies $|s^{-1}a|_k \leq |1|_k$. On the other hand, $|s^{-1}a|_i > |1|_i$ implies $|t|_i = |a^{-1}sb|_i \leq |a^{-1}s|_i < |1|_i$. Hence $t \in S_k$ and $at = sb$.

2.) Since S_k contains only units of B_k we have $RS_k^{-1} \subseteq B_k S_k^{-1} \subseteq B_k$.

Conversely, let $a \in B_k$, $a \neq 0$. By the lemma there exists $b \in R$ such that

$$|b|_i = |a|_i \quad \text{if} \quad |a|_i \leq |1|_i$$

$$|b|_i \leq |a^{-1}|_i \quad \text{if} \quad |a|_i > |1|_i.$$

Put $s = a^{-1}b$. Then $|a|_i \leq |1|_i$ implies $|s|_i = |a^{-1}b|_i = |1|_i$. This holds in particular for $i = k$ since $a \in B_i$. On the other hand, $|a|_i > |1|_i$ implies $|s|_i = |a^{-1}b|_i \leq |a^{-1}|_i < |1|_i$. Hence $s \in S_k$ and $a = bs^{-1} \in RS_k^{-1}$. Thus $B_k \subseteq RS_k^{-1}$.

THEOREM 7.5. Let I be a right ideal of $R = B_1 \cap \ldots \cap B_n$. Then we have

$$I = I_1 \cap \ldots \cap I_n$$

where $I_k = IS_k^{-1}$ are the right ideals of B_k which are generated by I.

PROOF : Obviously, I is contained in $I_1 \cap \ldots \cap I_n$. In order to show the inverse inclusion let $a \in I_1 \cap \ldots \cap I_n$, hence $a = a_k s_k^{-1}$, $a_k \in I$ and $s_k \in S_k$. Let $B_1, \ldots, B_i$ be the minimal valuation rings of the set $\{ B_1, \ldots, B_n \}$. Then $R = B_1 \cap \ldots \cap B_i$. Consider the right ideal C generated by $s_1, \ldots, s_i$. By 7.3 there exist $c_1, \ldots, c_i$ such that $|c_k|_k = |1|_k$ and $|c_k|_j < |1|_j$ for $j \neq k$ and $1 \leq j \leq i$. Then we have $c_k \in B_1 \cap \ldots \cap B_i = R$, hence

$$c = s_1 c_1 + \ldots + s_i c_i \in C.$$

By the principle of domination we have

$$|c|_k = |s_k c_k|_k = |1|_k \quad \text{for } k = 1, \ldots, i$$

Therefore c is invertible in R, hence C = R. Thus

$$1 = s_1 b_1 + \ldots + s_i b_i, \quad b_k \in R$$

implies

$$a = (as_1)b_1 + \ldots + (as_i)b_i$$

$$= a_1 b_1 + \ldots + a_i b_i \in I.$$

COROLLARY 7.6. Let I and J be right ideals of R such that $IS_k^{-1} = JS_k^{-1}$ for k = 1, ..., n. Then I = J.

PROOF : $I = IS_1^{-1} \cap \ldots \cap IS_n^{-1}$

$$= JS_1^{-1} \cap \ldots \cap JS_n^{-1} = J.$$

COROLLARY 7.7. The lattice of right ideals of R is distributive, i.e.

$$I \cap (J + L) = I \cap J + I \cap L$$

holds for right ideals I, J, L of R.

PROOF : Since the set of right ideals of a valuation ring is totally ordered by inclusion, distributivity holds for valuation rings. Therefore we have

$$(I \cap (J + L))S_k^{-1} = (I \cap J + I \cap L)S_k^{-1} \quad \text{for } k = 1, \ldots, n,$$

hence by 7.6

$$I \cap (J + L) = I \cap J + I \cap L.$$

For the proof of the next theorem we need a general form of the Chinese remainder theorem. As usual $a \equiv b \bmod I$ means $a - b \in I$.

PROPOSITION 7.8. Let R be a ring with a distributive right ideal lattice. Let $I_1, \ldots, I_n$ be right ideals of R and $a_1, \ldots, a_n \in R$. Then a system of congruences

(1) $$x \equiv a_1 \bmod I_1, \ \ldots, \ x \equiv a_n \bmod I_n$$

is solvable whenever, for all pairs of indices, the congruences

(2) $$x \equiv a_i \bmod I_i, \quad x \equiv a_k \bmod I_k$$

are solvable.

PROOF : We argue by induction on the number n of congruences. For n = 1 or n = 2 the assertions are obviously true. Let n > 2.

By (2) there exist $b_2, \ldots, b_n$ such that

(3) $$b_k \equiv a_1 \bmod I_1, \quad b_k \equiv a_k \bmod I_k$$

Consider now the system of n-1 congruences

(4) $$x \equiv b_2 \bmod (I_1 \cap I_2), \ldots, x \equiv b_n \bmod (I_1 \cap I_n).$$

First we will show that conditions of solvability are satisfied for (4). Let $i, k \geq 2$. By (2) and (3) we have

$$b_i - b_k = b_i - a_1 + a_1 - b_k \in I_1$$

$$b_i - b_k = b_i - a_i + a_i - a_k + a_k - b_k \in I_i + I_k,$$

hence

$$b_i - b_k \in I_1 \cap (I_i + I_k) = I_1 \cap I_i + I_1 \cap I_k$$

by the distributivity of the right ideals. From this follows the solvability of

(5) $$x \equiv b_i \bmod (I_1 \cap I_i), \quad x \equiv b_k \bmod (I_1 \cap I_k),$$

namely, if $b_i - b_k = u + v$ with $u \in I_1 \cap I_i$, $v \in I_1 \cap I_k$ then $x = b_i - u$ is a solution of (5). From the solvability of (5) we conclude by induction that (4) is solvable. Let x denote a solution of (4). Then we have

$$x \equiv b_2 \equiv a_1 \bmod I_1$$

$$x \equiv b_k \equiv a_k \bmod I_k \quad \text{for } k = 2, \ldots, n.$$

Hence x is also a solution of (1).

PROPOSITION 7.9. Let $a_1, \ldots, a_k \in K$. Then there exists $b \neq 0$ in R such that $ba_i \in R$ for $i = 1, \ldots, k$.

PROOF : Since K is the quotient field of R by 7.2 there exists $c \neq 0$ in R such that $ca_1 \in R$. By induction there exists $d \neq 0$ in R such that $dca_i \in R$ for $i = 2, \ldots, k$, hence $b = dc$ has the property required above.

THEOREM 7.10. Let $| \ |_1, \ldots, | \ |_n$ be valuations of K with the value sets $W_1, \ldots, W_n$. For given $a_i \in K$ and $\varepsilon_i \in W_i^{\times}$ there exists $a \in K$ such that

$$|a - a_i|_i \leq \varepsilon_i \quad \text{for } i = 1, \ldots, n.$$

whenever for each pair of indices there exists $a_{ik} \in K$ such that

$$|a_{ik} - a_i|_i \leq \varepsilon_i \quad \text{and} \quad |a_{ik} - a_k|_k \leq \varepsilon_k .$$

PROOF : By 7.9 we find $b \neq 0$ in R such that $ba_{ik} \in R$ and $ba_i \in R$ for all i and k. Define the right ideals of R

$$I_i = \{ c \in R \mid |b^{-1}c|_i \leq \varepsilon_i \}.$$

Then the element $ba_{ik} \in R$ is a solution of

$$x \equiv ba_i \bmod I_i , \quad x \equiv ba_k \bmod I_k .$$

By 7.7 R possesses a distributive right ideal lattice. We may therefore apply the Chinese remainder theorem on R. Hence there exists $c \in R$ such that $c \equiv ba_i \bmod I_i$ for all $i = 1, \ldots , n$. Now $a = b^{-1}c$ satisfies the assertion of the theorem.

REMARK : It should be emphasized that theorem 7.10 is valid for general valuations. In the next section we shall require additional conditions on the valuations such that the solvability conditions are satisfied for all $a_i \in K$ and all $\varepsilon_i \in W_i^{\times}$. By 7.10 it is sufficient to consider only two valuations.

8. Independence of valuations

Two valuations $| \ |_1 : K \longrightarrow W_1$ and $| \ |_2 : K \longrightarrow W_2$ are called **independent** when for any $\varepsilon_1 \in W_1^\times$, $\varepsilon_2 \in W_2^\times$ there exists an element a of K such that

$$|a|_1 = \varepsilon_1 \quad \text{and} \quad |a|_2 = \varepsilon_2 .$$

PROPOSITION 8.1. For two valuations $| \ |_i : K \longrightarrow W_i$ with the valuation rings B_i, i = 1, 2, the following assertions are equivalent

(1) $| \ |_1$ and $| \ |_2$ are independent.

(2) For any $\varepsilon_1 \in W_1^\times$ there exists $a \in K$ such that

$$|a|_1 = \varepsilon_1 \quad \text{and} \quad |a|_2 = |1|_2 .$$

(3) For any $\varepsilon_1 \in W_1^\times$ there exists $a \in K$ such that

$$|a|_1 \leq \varepsilon_1 \quad \text{and} \quad |1 - a|_2 \leq \varepsilon_2 .$$

(4) For any $a_i \in K$, $\varepsilon_i \in W_i^\times$, i = 1, 2, there exists $a \in K$ such that

$$|a - a_1|_1 \leq \varepsilon_1 \quad \text{and} \quad |a - a_2|_2 \leq \varepsilon_2$$

PROOF : (1) ⇒ (2) is immediate.

(2) ⇒ (3). We may assume $\varepsilon_1 < |1|_1$. Choose $b_1, b_2 \in K$ such that $|b_1|_1 = \varepsilon_1$ and $|b_2|_2 = \varepsilon_2$. (This is possible since valuations are surjections by definition.) $I_1 = b_1B_1 \cap B_2$ and $I_2 = B_1 \cap b_2B_2$ are right ideals of $R = B_1 \cap B_2$. In order to show $I_1 + I_2 = R$ suppose that $I_1 + I_2$ is a proper right ideal of R. By 7.5 it is contained in $M_1 \cap B_2$ or $B_1 \cap M_2$ where M_1 and M_2 are the maximal ideals of the valuation rings, say $I_1 + I_2 \subseteq M_1 \cap B_2$.

Let $S = \{ s \in R \mid |s|_1 = |1|_1 \}$. We shall prove $I_2S^{-1} = B_1$. By (2) there exists $c \in K$ such that

$$|c|_1 = |b_2^{-1}|_1 \quad \text{and} \quad |c|_2 = |1|_2 .$$

Put $s = b_2c$. Then $|s|_1 = |1|_1$ and $s \in R$, hence $s \in S$. Further, $s \in B_1 \cap b_2B_2 = I_2$. Therefore, we have $I_2S^{-1} = B_1$, since $1 = ss^{-1} \in I_2S^{-1}$.

From $I_2 \subseteq I_1 + I_2 \subseteq M_1 \cap B_2$ it follows by localizing with S

$$B_1 = I_2 S^{-1} \subseteq (M_1 \cap B_2)S^{-1} = M_1$$

a contradiction. Thus $I_1 + I_2 = R$. If $1 = a + b$, $a \in I_1$, $b \in I_2$, then $a \in b_1 B_1$ and $1 - a \in b_2 B_2$, hence $|a|_1 \leq |b_1|_1 = \varepsilon_1$ and $|1 - a|_2 \leq |b_2|_2 = \varepsilon_2$.

(3) ⇒ (4). If $a_1 = a_2$ put $a = a_1 = a_2$. Assume $a_1 \neq a_2$. Let b_1, $b_2 \in K$ such that $|b_1|_1 = \varepsilon_1$ and $|b_2|_2 = \varepsilon_2$. By (3) there exists $b \in K$ such that

$$|b|_1 \leq |(a_1 - a_2)^{-1} b_1|_1 \quad \text{and} \quad |1 - b|_2 \leq |(a_1 - a_2)^{-1} b_2|_2$$

hence

$$|(a_1 - a_2)b|_1 \leq \varepsilon_1 \quad \text{and} \quad |(a_1 - a_2)(1 - b)|_2 \leq \varepsilon_2 .$$

Then $a = a_1 - (a_1 - a_2)b$ satisfies the conditions required.

(4) ⇒ (1). Choose $a_i \in K$ with $|a_i|_i = \varepsilon_i$ and $\varepsilon_i' \in W_i^{\times}$ such that $\varepsilon_i' < \varepsilon_i$ for $i = 1, 2$. By (4) there exists $a \in K$ such that $|a - a_i|_i \leq \varepsilon_i'$. Then we have by the principle of domination

$$|a|_i = |a_i + (a - a_i)|_i = |a_i|_i = \varepsilon_i .$$

As a corollary to 7.10 we obtain

THEOREM 8.2. Let $| \; |_i : K \longrightarrow W_i$ be valuations of K and $a_i \in K$, $\varepsilon_i \in W_i^{\times}$, $i = 1, \ldots, n$. If the valuations are pairwise independent then there exists an element a of K such that

$$|a - a_i|_i \leq \varepsilon_i \quad \text{for} \quad i = 1, \ldots, n.$$

A weaker form of independence we obtain by considering the induced topologies.

PROPOSITION 8.3. Independent valuations induce distinct topologies.

PROOF : Let B_1, B_2 be the valuation rings of the valuations. Suppose that the topologies generated by B_1 and B_2 are equal. Then the maximal ideal M_2 of B_2 is a neighbourhood of 0 in the topology generated by B_1. Hence there exists $a \in K^{\times}$ such that $aB_1 \subseteq M_2$. Since the valuations are independent there exists $b \in K$ such that

$$|b|_1 = |a|_1 \quad \text{and} \quad |b|_2 = |1|_2.$$

Then $a^{-1}b \in B_1$, hence $b = a(a^{-1}b) \in aB_1 \subseteq M_2$ contrary to $|b|_2 = |1|_2$.

Conjugate valuations are never independent. Namely, when $|\ |: K \longrightarrow W$ is a valuation with the valuation ring B then the valuation $|\ |' : K \longrightarrow W$ defined by

$$|x|' = |xa|$$

corresponds to the valuation ring aBa^{-1}. Obviously, both valuations are dependent. By 6.3 there exists for each valuation, which is is not a V-valuation, a conjugate valuation inducing a different topology. Hence, the converse of 8.3 is not valid in general, but it is true for V-valuations as we now will show.

PROPOSITION 8.4. Two V-valuations are independent if they induce distinct topologies.

PROOF : Let $U_i = \{ a \in K \mid |a|_i \leq \varepsilon_i \}$ be neighbourhoods of 0 of the given V-valuations $|\ |_i$, $i = 1, 2$. When the induced topologies are distinct, then by 5.5 $U_1 \cap (1 + U_2) \neq \emptyset$. Each element of the intersection satisfies the condition (3) of 8.1, hence the valuations are independent.

THEOREM 8.5. Let $|\ |_i : K \longrightarrow W_i$, $i = 1, \ldots, n$, be V-valuations of K inducing distinct topologies. If $a_i \in K$, $\varepsilon_i \in W_i^{\times}$ are given then there exists $a \in K$ such that

$$|a - a_i|_i \leq \varepsilon_i \quad \text{for} \quad i = 1, \ldots, n.$$

REMARK : This theorem holds more generally for V-topologies which are not necessarily induced by valuations (where $|a - a_i|_i \leq \varepsilon_i$ is replaced by $a \in a_i + U_i$). This was shown by A. L. Stone for commutative fields using non-standard methods (Stone [1]) and in general by H. Weber [1].

THEOREM 8.6. Let $| \ |_1$ be a V-valuation and $| \ |_2$ an arbitrary valuation of K. If the topology induced by $| \ |_1$ is not finer than the topology induced by $| \ |_2$ then $| \ |_1$ and $| \ |_2$ are independent.

PROOF : We prove that for any $\varepsilon_1 \in W_1^x$ there exists $c \in K$ such that $|c|_1 = \varepsilon_1$ and $|c|_2 = |1|_2$. The assertion then follows from 8.1. We distinguish two cases.

1.) Let $\varepsilon_1 < |1|_1$. Choose $a \in K$ such that $|a|_1 = \varepsilon_1$. By the assumption on the topologies aB_1 is not contained in B_2 (otherwise, the topology generated by B_1 would be finer than the topology generated by B_2). Hence there exists $b \in B_1$ such that $ab \notin B_2$. Put

$$d = a \qquad \text{if} \qquad |a|_2 > |1|_2,$$

$$d = a(1 + b) \qquad \text{if } |ab|_1 < |a|_1 \text{ and } |a|_2 \leq |1|_2,$$

$$d = ab \qquad \text{if } |ab|_1 = |a|_1 \text{ and } |a|_2 \leq |1|_2.$$

In all cases we have $|d|_1 = \varepsilon_1 < |1|_1$ and $|d|_2 > |1|_2$. Then $1 + d$ is a unit in B_1 and $1 + d^{-1}$ a unit in B_2. We put $c = d(1 + d)^{-1} = (1 + d^{-1})^{-1}$. Then $|c|_1 = \varepsilon_1$ and $|c|_2 = |1|_2$.

2.) Let $\varepsilon_1 \geq |1|_1$. Since $| \ |_1$ is a V-valuation by 6.5 there exists $\varepsilon_1' \in W_1^x$, $\varepsilon_1' < |1|_1$, such that

(*) $\quad |a| \leq \varepsilon_1'$ implies $\quad |a^{-1}|_1 > \varepsilon_1$ for all $a \in K^x$.

As was shown in the first case, there exists $v \in K$ such that $|v|_1 = \varepsilon_1'$ and $|v|_2 = |1|_2$. Choose $u \in K$ such that $|u|_1 = \varepsilon_1$. By (*) we have $|v^{-1}|_1 > \varepsilon_1 = |u|_1$, hence $|vu|_1 < |1|_1$. Again, we apply the first case and obtain an element w such that $|w|_1 = |vu|_1$ and $|w|_2 = |1|_2$. Now put $c = v^{-1}w$. Then $|c|_1 = |u|_1 = \varepsilon_1$ and $|c|_2 = |v^{-1}w|_2 = |v^{-1}|_2 = |1|_2$.

Two valuation rings of K are said to be **comaximal** if there does not exist a valuation ring $B \neq K$ containing both. By 1.7 it is equivalent to say that the valuation rings possess no common completely prime ideal $P \neq 0$. Two valuations are called comaximal if their valuation rings are comaximal. If two valuations induce distinct topologies, by 6.2 they are comaximal.

THEOREM 8.7. Let $| \ |_1$ be an invariant valuation and $| \ |_2$ an arbitrary valuation of K. If they are comaximal then they are independent.

PROOF : Let B_1 and B_2 be the associated valuation rings and M_1 and M_2 their maximal ideals. Consider

$$P = \{ a \in K \mid aB_1 \subseteq M_2 \}.$$

Since $1 \in B_1$ we have $P \subseteq M_2$. Further $P \subseteq M_1$, namely $a \in P$, $a \notin M_1$ would imply $a^{-1} \in B_1$, hence $1 = a\, a^{-1} \in aB_1 \subseteq M_2$, a contradiction.

We show that P satisfies the condition (P) of 1.6. Let a, b $\in K$ and $ab \in P$, $b \notin P$. Then there exists $c \in B_1$ such that $bc \notin M_2$, hence $(bc)^{-1} \in B_2$. By the invariance of B_1 we obtain

$$aB_1 = ab\, c\, ((bc)^{-1}B_1(bc))\, (bc)^{-1} \subseteq abB_1B_2 \subseteq M_2B_2 \subseteq M_2,$$

hence $a \in P$.

By 1.6 P is a completely prime ideal of B_1 and B_2. Since B_1 and B_2 are comaximal we have $P = 0$. This means that the topology induced by $|\ |_1$ is not finer than the topology induced by $|\ |_2$. Further, $|\ |_1$ is a V-valuation. Applying 8.6 we obtain the independence of $|\ |_1$ and $|\ |_2$.

REMARK : In section 4 a non-invariant valuation ring B was constructed with only one non-trival two-sided ideal. By 6.4 the valuation belonging to B is not a V-valuation, by 6.3 there exists a conjugate valuation ring which generates a topology being distinct from the topology generated by B. But conjugate valuations are not independent. This shows that 8.5 is not valid for general valuations.

Dubrovin [2] has shown that there exist non-invariant maximal V-valuation rings. Two different conjugates of such a valuation ring are comaximal, but they generate the same topology. Therefore, 8.7 is not valid for V-valuations.

Let us collect the results of this section. We studied the relations of the following statements

(1) $|\ |_1$ and $|\ |_2$ are independent.

(2) $|\ |_1$ and $|\ |_2$ induce distinct topologies.

(3) $|\ |_1$ and $|\ |_2$ are comaximal.

The implications (1) $\Rightarrow$ (2) $\Rightarrow$ (3) are valid for arbitrary valuations. If one of the valuation is invariant then all statements are equivalent. Neither (2) $\Rightarrow$ (1) nor (3) $\Rightarrow$ (2) holds in general. For V-valuations (1) and (2) are equivalent.

EXERCISES

1.) Show that a locally bounded topology which is induced by a valuation is a V-topology.

2.) Show that a valuation $|\ |: K \longrightarrow W$ is a V-valuation if and only if the following condition holds: For every $\varepsilon \in W^{\times}$ there exists $\varepsilon' \in W^{\times}$ such that $|a| > \varepsilon'$ implies $|a^{-1}| \leq \varepsilon$ for all $a \in K^{\times}$.

3.) Let $|\ |_1, \ldots, |\ |_n$ be pairwise independent valuations of K. Show that for given $\varepsilon_i \in W_i^{\times}$, $i = 1, \ldots, n$, there exists $a \in K$ such that $|a|_i = \varepsilon_i$ for $i = 1, \ldots, n$.

4.) Let B_1, B_2 be valuation rings of K with the maximal ideals M_1 and M_2. Then $P = \{ a \in K \mid aB_1 \subseteq M_2 \}$ is a proper right ideal of B_1 and a proper left ideal of B_2. If B_1 and B_2 are topologically equivalent then $P \neq 0$.

5.) Let $|\ |_1$ and $|\ |_2$ be two valuations of K with the valuation rings B_1 and B_2. Prove that they are independent if and only if $B_1 a B_2 = K$ for all $a \in K^{\times}$.

6.) Let $|\ |_1$ and $|\ |_2$ be independent valuations of K and $|\ |_2'$ a valuation which is conjugate to $|\ |_2$. Then $|\ |_1$ and $|\ |_2'$ are independent.

7.) A valuation is said to be infinite if its valuation ring is not contained in a maximal valuation ring. Show that an infinite valuation is a V-valuation.

8.) Let $|\ |_1$ be an infinite valuation, $|\ |_2$ an arbitrary valuation. Show that the valuations are independent if and only if they are comaximal.

9.) An element a of a valued field K is said to be analytically nilpotent when $\lim a^n = 0$ in the topology induced by the valuation. Let N be the set of all analytically nilpotent elements of K. Show $N = 0$ if the valuation is infinite.

CHAPTER 3 VALUED VECTOR SPACES

Valued vector spaces are special kinds of normed vector spaces. The main result of this chapter is theorem 11.1. This theorem is the basis of the determination of the union and intersection sets in the projective Hjelmslev spaces considered in the next chapter.

9. Normed vector spaces

Let K be a field and $|\ | : K \longrightarrow W$ a valuation of K. A (left) vector space T over K is called a **normed vector space** if there exists a norm $T \longrightarrow W$, $x \longmapsto \|x\|$, satisfying

(N1) $\|x\| = 0 \iff x = 0$,

(N2) $\|x + y\| \leq \mathrm{Max}(\|x\|, \|y\|)$,

(N3) $\|ax\| = \tilde{a}\|x\|$

for $x, y \in T$, $a \in K$. Further, we assume that T is not the nullspace.

There are some immediate consequences of these axioms which are proved similarly to the corresponding statements for valuations.

(1) For all $x \in T$ we have $\|x\| = \|-x\|$.

(2) The **principle of domination** holds: If $\|x\| \neq \|y\|$ then $\|x + y\| = \mathrm{Max}(\|x\|, \|y\|)$.

(3) Let x be a vector of T. Since a valuation is a surjective mapping there exists an element $a \in K$ such that $\|x\| = |a|$. From this follows that for $x \in T$ there exist $a \in K$ and $x' \in T$ such that

$$x = a\, x' \quad \text{and} \quad \|x'\| = |1|.$$

We say that we normalize the vector x.

(4) The set

$$T_1 = \{\ x \in T \mid \|x\| \leq |1|\ \}$$

corresponds to the valuation ring B of K. T_1 is a B-submodule of T.

Let I be a proper two-sided ideal of B. We define

$$T_I = \{ x \in T \mid \|x\| = |a| \text{ for some } a \in I \}.$$

PROPOSITION 9.1. T_I is a B-submodule of T_1 and we have $T_I = IT_1$.

PROOF : We first show that $x \in T_I$ and $\|y\| \leq \|x\|$ imply $y \in T_I$. Let $\|x\| = |a|$, $a \in I$, and $\|y\| = |b|$. Since $|b| \leq |a|$ we have $a^{-1}b \in B$, hence $b = a(a^{-1}b) \in I$, hence $y \in T_I$.

Let $x, y \in T_I$. Then $\|x + y\| \leq \text{Max}(\|x\|, \|y\|)$, hence $x + y \in T_I$ by the rule just proved. Further, let $a \in B$, $x \in T_I$. Then, there exists $b \in I$ such that $\|x\| = |b|$, thus

$$\|ax\| = \tilde{a}\|x\| = \tilde{a}|b| = |ab|.$$

Since $ab \in I$ we have $ax \in T_I$. Thus T_I is a B-submodule of T_1.

We show that, $x \in T_I$ and $\|x\| = |a|$ imply $a \in I$. Namely, let $\|x\| = |b|$, $b \in I$. Then $|a| = |b|$, hence $b^{-1}a \in B$, hence $a = b(b^{-1}a) \in I$.

We are now ready to prove $T_I = IT_1$. Every vector $x \in T$ may be written as $x = ax'$ where $\|x'\| = |1|$. If $x \in T_I$ we have $\|x\| = \tilde{a}\|x'\| = |a|$, hence $a \in I$ and therefore $x = ax' \in IT_1$. Conversely, let $x = ax' \in IT_1$ with $a \in I$, $x' \in T_1$. If $\|x'\| = |b|$, $b \in B$, we have

$$\|x\| = \tilde{a}\|x'\| = \tilde{a}|b| = |ab|.$$

Since $ab \in I$ we obtain $x \in T_I$.

In particular, when $I = M$ is the maximal ideal of B we have

$$T_M = \{ x \in T \mid \|x\| < |1| \} = M\, T_1.$$

Therefore the factor module

$$\overline{T} = T_1 / T_M$$

is annihilated by M. Hence $\overline{T}$ may be regarded as a vector space over the residue class field

$$\overline{K} = B/M.$$

$\overline{T}$ is called the **residue class space**. If

$$x \longmapsto \bar{x} \quad \text{and} \quad a \longmapsto \bar{a}$$

are the canonical homomorphisms $T_1 \longrightarrow \overline{T}$ and $B \longrightarrow \overline{K}$ we obtain

$$\overline{ax} = \bar{a}\,\bar{x}$$

and

$$\bar{x} \neq 0 \quad \Longleftrightarrow \quad \|x\| = |1|$$

for $a \in B$, $x \in T_1$.

We call subspaces X_i of T, where i runs through an index set J, **normally independent** if

$$\|x\| = \operatorname{Max}(\|x_i\|) \quad \text{for } x = \sum_{i \in J} x_i \in \sum_{i \in J} X_i .$$

We call the vectors $x_i \neq 0$ **normally independent** when the subspaces $X_i = \langle x_i \rangle$ generated by x_i are normally independent.

A basis $\{x_i\}$ of a subspace X of T is called a **normal basis** if the vectors x_i are normally independent and $\|x_i\| = |1|$.

PROPOSITION 9.2. The vectors $x_i \in T$, $\|x_i\| = |1|$, are normally independent if and only if the vectors $\bar{x}_i \in \overline{T}$ are linearly independent over $\overline{K}$.

PROOF : We may assume that the numbers of vectors is finite, because any linear expression $x = \Sigma\, a_i x_i$ contains only a finite number of non-zero summands.

Let therefore $\bar{x}_1, \ldots, \bar{x}_n$ be linearly independent over $\overline{K}$,

$$x = \sum_{i=1}^{n} a_i x_i ,$$

and say

$$|a_1| = \operatorname*{Max}_{i=1...n} (|a_i|) = \operatorname*{Max}_{i=1...n} (\|a_i x_i\|)$$

Then we have

$$a_1^{-1}x = x_1 + \sum_{i=2}^{n} b_i x_i \in T_1$$

with $b_i = a_1^{-1}a_i \in B$. By applying the residue class homomorphism we obtain

$$\overline{a_1^{-1}x} = \overline{x}_1 + \sum_{i=2}^{n} \overline{b}_i \overline{x}_i \neq 0$$

since $\overline{x}_1, \ldots, \overline{x}_n$ are linearly independent. Hence $\|a_1^{-1}x\| = |1|$ and therefore

$$\|x\| = |a_1| = \operatorname*{Max}_{i}(\|a_i x_i\|).$$

Conversely, let $\overline{x}_1, \ldots, \overline{x}_n$ be linearly dependent over $\overline{K}$. Then there is a non-trivial linear relation

$$\sum_{i=1}^{n} \overline{a}_i \overline{x}_i = 0.$$

Put $x = \Sigma\, a_i x_i$. Then $\overline{x} = 0$, hence $\|x\| < |1|$. But for at least one index we have $\|a_i x_i\| = |1|$, hence $\|\Sigma\, a_i x_i\| < \operatorname{Max}(\|a_i x_i\|)$. This means that $x_1, \ldots, x_n$ are not normally independent.

Let X be a subspace of T. Put

$$X_1 = \{ x \in X \mid \|x\| \leq |1| \}$$

and

$$\varphi(X) = \{ \overline{x} \in \overline{T} \mid x \in X_1 \}.$$

Then $\varphi(X)$ is a subspace of $\overline{T}$ and φ maps the lattice of subspaces of T onto the lattice of subspaces of $\overline{T}$. It should be mentioned that φ preserves inclusion but that φ is not a lattice homomorphism.

PROPOSITION 9.3. Let Y be a subspace of T, $\overline{X}$ a subspace of $\overline{T}$ with $\overline{X} \subseteq \varphi(Y)$. Then there exists a subspace X of Y such that $\varphi(X) = \overline{X}$ and X has the same dimension as $\overline{X}$.

PROOF : Let $\{\overline{x}_i\}$ be a basis of $\overline{X}$. Since $\overline{X} \subseteq \varphi(Y)$ we can find $x_i \in Y$ being mapped onto $\overline{x}_i$ by the residue class homomorphism. The vectors x_i are linearly independent. Namely, let $\Sigma\, a_i x_i = 0$ be a non-trivial linear relation. We may assume $\mathrm{Max}(|a_i|) = |1|$, otherwise, we multiply with a suitable element of K. Now applying the residue class homomorphism we get a contradiction to the linear independence of the vectors $\overline{x}_i$.

Let X be the subspace spanned by the vectors x_i. Obviously, $X \subseteq Y$ and $\dim_K X = \dim_{\overline{K}} \overline{X}$. Since $\overline{x}_i \in \varphi(X)$ we have $\overline{X} \subseteq \varphi(X)$. Let $x = \Sigma\, a_i x_i \in X_1$. Since the vectors x_i are normally independent we have

$$\|x\| = \mathrm{Max}(\|a_i x_i\|) = \mathrm{Max}(|a_i|) \leq |1|,$$

hence $a_i \in B$ and therefore

$$\overline{x} = \Sigma\, \overline{a}_i \overline{x}_i \in \overline{X}.$$

Thus $\varphi(X) = \overline{X}$.

COROLLARY 9.4. For any subspace X of T we have $\dim_{\overline{K}} \varphi(X) \leq \dim_K X$.

PROOF : By 9.3 there exists a subspace of X of the same dimension as $\varphi(X)$.

10. Valued vector spaces

A normed vector space T is called a **valued vector space** if for every $y \in T$ and every finite dimensional subspace X of T there exists an element $x_0 \in X$ such that

$$\|y - x_0\| \leq \|y - x\| \quad \text{for all } x \in X.$$

x_0 is called a **best approximation** of y in X.

THEOREM 10.1. For a normed vector space T the following assertions are equivalent

(1) T is a valued vector space,

(2) If X is a finite dimensional subspace of T, then $X_1 = X \cap T_1$ is a finitely generated B-module,

(3) If X is a finite dimensional subspace of T then $\dim_K X = \dim_{\overline{K}} \varphi(X)$.

PROOF : (1) ⇒ (2). We argue by induction on n = dim X. If dim X = 0 then $X_1 = 0$ is finitely generated. Let dim Y = n+1 and $Y = X + \langle y \rangle$ with dim X = n. By induction X_1 is finitely generated, say

$$X_1 = \sum_{i=1}^{k} B\, x_i$$

Since T is a valued vector space there exists $x_0' \in X$ such that

$$\|y - x_0'\| \leq \|y - x\| \quad \text{for all } x \in X.$$

Since $y \notin X$ we have $y - x_0' \neq 0$. Write $y - x_0' = a\, x_0$ with $\|x_0\| = |1|$, $a \in K^x$. Then we obtain

$$|1| = \|x_0\| \leq \|x_0 + x\| \quad \text{for all } x \in X$$

and

$$Y = \langle x_0 \rangle + X.$$

We will show that $x_0, x_1, \ldots, x_k$ generate Y_1. Let $z \in Y$, $z \notin X$, $\|z\| \leq |1|$, and

$$z = a_0 x_0 + x \quad \text{with } x \in X,\ a_0 \in K^{\times}.$$

Since

$$|1| \geq \|z\| = \tilde{a}_0 \|x_0 + a_0^{-1}x\| \geq |a_0|$$

we obtain $a_0 \in B$. This implies $x = z - a_0 x_0 \in X_1$, hence

$$x = \sum_{i=1}^{k} a_i x_i, \quad a_i \in B$$

and therefore

$$z = \sum_{i=0}^{k} a_i x_i \in \sum_{i=0}^{k} B\, x_i.$$

It follows

$$Y_1 = \sum_{i=0}^{k} B\, x_i.$$

(2) ⇒ (3). By 9.3 there exists a subspace X' contained in X such that $\varphi(X') = \varphi(X)$ and $\dim_K X' = \dim_{\overline{K}} \varphi(X)$. We have to show $X = X'$.

We consider the exact sequence of B-modules

$$0 \longrightarrow X'_1 \xrightarrow{\ i\ } X_1 \xrightarrow{\ j\ } W \longrightarrow 0,$$

where i is the embedding and W is the factor module X_1/X'_1. In order to show $W \subseteq M\,W$, we choose for an element $w \in W$ an element $x \in X_1$ such that $j(x) = w$. Since $\varphi(X) = \varphi(X')$ there exists an element $x' \in X'_1$ such that $x - x' \in T_M$. Write $x - x' = a\,y$ with $\|y\| = |1|$, $a \in K$. Since $x - x' \in X$ we may assume $y \in X_1$. Further, since

$$|1| > \|x - x'\| = \tilde{a}\|y\| = |a|$$

we obtain $a \in M$ and

$$w = j(x) = j(x' + ay) = aj(y) \in M\,W.$$

Now (2) implies that X_1 and therefore W is finitely generated. By the lemma of Nakayama we have $W = 0$ and thereby $X'_1 = X_1$ which implies $X' = X$.

(3) $\Rightarrow$ (1). If $y \in X$ then obviously $x_0 = y$ is a best approximation of y. We may therefore assume $y \notin X$. Let $Z = X + \langle y \rangle$. Then $\dim_K Z = \dim_K X + 1$ and by (3) $\dim_{\overline{K}} \varphi(Z) = \dim_{\overline{K}} \varphi(X) + 1$. Hence there exists $z \in Z$ which is normally independent to X such that $Z = X + \langle z \rangle$. Let $y = x_0 + az$, $x_0 \in X$, $a \in K$. Then we obtain

$$\|y - x\| = \|y - x_0 + x_0 - x\| = \mathrm{Max}(\|y - x_0\|, \|x_0 - x\|),$$

hence

$$\|y - x_0\| \leq \|y - x\| \quad \text{for all } x \in X.$$

Thus T is a valued vector space.

The following theorem gives a sufficient condition under which a normed vector space is a valued vector space. In the finite dimensional case this condition is also necessary.

THEOREM 10.2. Let T be a normed vector space possessing a normal basis. Then T is a valued vector space.

PROOF : We want to prove condition (3) of 10.1. Let X be a finite dimensional subspace of T and $\{x_1, \ldots, x_n\}$ a basis of X. We may assume $\|x_i\| = |1|$. If $\{y_i\}$ is a normal basis of T we have

$$x_i = \Sigma \; a_{ik} \, y_k, \quad i = 1, \ldots, n$$

Since in each expression there are a finite number of non-zero coefficients, only a finite number of basis vectors y_i occur in the expressions, say the first m. Therefore we obtain

$$x_i = \sum_{k=1}^{m} a_{ik} \, y_k, \quad i = 1, \ldots, n$$

Since the vectors y_i are normally independent we have

$$\|x_i\| = \underset{k=1\ldots m}{\mathrm{Max}} \; (|a_{ik}|) = |1|$$

Hence the coefficients a_{ik} are elements of B and for every i there exists at least one coefficient a_{ik} with value $|1|$. We may assume, perhaps by changing the indices of the vectors y_i, $|a_{11}| = |1|$.

We now consider the vectors

$$x_2 - a_{21}\, a_{11}^{-1}\, x_1,\ \dots\ ,\ x_n - a_{n1}\, a_{11}^{-1}\, x_1$$

and their representations in the basis $\{y_i\}$. The coefficient of y_1 is always zero. Further, they form together with x_1 a basis of X. When we normalize the vectors and proceed in the same manner, we obtain finally a basis of X in triangle form

$$x_i' = \sum_{k=1}^{m} a_{ik}'\, y_k, \qquad i = 1,\ \dots\ , n$$

such that

$$a_{ik}' = 0 \quad \text{for} \quad i < k \quad \text{and} \quad |a_{ii}| = |1|.$$

Now the images of x_i' in the residue class space are linearly independent. Therefore $\dim_K X = \dim_{\overline{K}} \varphi(X)$.

Consider for a moment the finite dimensional case. The vector space $T = K^n$ with the cubical norm

$$\|x\| = \|(a_1,\ \dots\ ,\ a_n)\| = \operatorname{Max}_i(|a_i|)$$

possesses the normal basis $\{(1,\dots,0),\ \dots\ ,\ (0,\dots,1)\}$ and is therefore by 10.2 a valued vector space. On the other hand, let T be an arbitrary valued vector space of finite dimension, say n. The residue class space $\overline{T}$ has dimension n. Let $\{x_1,\ \dots\ ,\ x_n\}$ be a basis of T such that the images of the vectors x_i under the residue class homomorphism form a basis of $\overline{T}$. Therefore, $\{x_1, \dots\ , x_n\}$ is a normal basis of T. If $x = \Sigma\, a_i x_i$ is an element of T we have

$$\|x\| = \|\Sigma a_i x_i\| = \operatorname{Max}\ (\|a_i x_i\|) = \operatorname{Max}\ (|a_i|).$$

This equation shows that there exists a basis of T such that the norm of T is the cubical norm with respect to this basis. Hence, in the finite dimensional case vector spaces with cubical norm and valued vector spaces are equivalent notions.

Examples of normed vector spaces which are not valued are easily found in algebraic number theory. Let T be a finite extension of the rational number field. A valuation of T can be interpreted as norm and T is a normed vector space if the valuation is unramified. By 10.1 T is valued if the valuation is inertial.

11. Finitely generated B-modules

THEOREM 11.1. Let U be a finitely generated B-submodule of a valued vector space T. Then there exists a system of normally independent generators.

PROOF : Let $X = \langle U \rangle$ be the subspace of T spanned by U. Since U is finitely generated the dimension of X is finite, say n. By 10.1 the mapping φ preserves the dimension, hence

$$n = \dim_K X = \dim_{\overline{K}} \varphi(X)$$

where $\overline{K}$ is the residue class field.

Let $S = \{x_1, \ldots, x_k\}$ be a system of generators of U. We define

$$\overline{S} = \sum_{i=1}^{k} \varphi(\langle x_i \rangle)$$

where $\langle x_i \rangle$ is the subspace generated by x_i.

Since $\overline{S}$ is a subspace of $\varphi(X)$ we have $\dim_{\overline{K}} \overline{S} \leq n$.

From the set of all systems of generators of U we choose one system $S = \{x_1, \ldots, x_k\}$ for which $\dim_{\overline{K}} \overline{S}$ is maximal. We claim that $\varphi(X) = \overline{S}$. Assuming $\overline{S}$ is a proper subspace of $\varphi(X)$ there exists a vector

$$x = \sum_{i=1}^{k} a_i x_i \in X_1, \quad a_i \in K$$

such that $\bar{x} \notin \overline{S}$. Let $a \in K$ such that $|a| = \mathrm{Max}\,(|a_i|)$. Then

$$x_0 = a^{-1}x = \sum_{i=1}^{k} b_i x_i$$

is contained in U since $b_i = a^{-1}a_i \in B$. The system $S' = S \cup \{x_0\}$ generates U. Further, $\bar{x} \in \overline{S'}$, but $\bar{x} \notin \overline{S}$. Hence $\overline{S}$ is a proper subspace of $\overline{S'}$, contrary to the maximal dimension of $\overline{S}$.

In the following we shall consider only systems $S = \{x_1, \ldots, x_k\}$ of generators of U satisfying

$$(1) \qquad \sum_{i=1}^{k} \varphi(\langle x_i \rangle) = \varphi(X).$$

Clearly $n \leq k$. Now we shall show that the number of generators may be reduced if $n < k$.

Let $n < k$. Then the k vectors x_i are linearly dependent over K. We consider a non-trivial linear relation

$$(2) \qquad \sum_{i=1}^{k} a_i x_i = 0, \quad a_i \in K.$$

Let l be the number of the non-zero coefficients a_i in (2). If $l = 1$ one generator is 0. Then the others form a system of $k-1$ generators of U.

Let $l > 1$. We consider the norms $\|a_i x_i\|$ of the summands in (2). We may assume that their maximum is $|1|$. (Otherwise multiply (2) by c^{-1} where $|c|$ is the maximum.) Further, by a suitable choice of the indices we may assume that

$$\|a_i x_i\| = |1| \quad \text{and} \quad |a_1| \geq |a_i| \text{ for } i = 1, \ldots, r$$

$$\|a_i x_i\| < |1| \qquad \text{for } i = r+1, \ldots, k.$$

By the principle of domination there are two summands with maximal norm, thus $r \geq 2$.

Applying the residue class homomorphism onto (2) we obtain

$$(3) \qquad \sum_{i=1}^{r} \overline{a_i x_i} = 0$$

We now replace x_1 by

$$x_1' = x_1 + a_1^{-1} \sum_{i=2}^{r} a_i x_i$$

Since $a_1^{-1} a_i \in B$ we have $x_1' \in U$. Obviously, $S' = \{x_1', x_2, \ldots, x_k\}$ generates U.

Since $\varphi(\langle x_i \rangle) = \langle \overline{a_i x_i} \rangle$ for $i = 1, \ldots, r$ we have

$$\varphi(\langle x_1 \rangle) \subseteq \sum_{i=2}^{r} \varphi(\langle x_i \rangle)$$

by (3), hence (1) may be shorten to

$$\sum_{i=2}^{k} \varphi(\langle x_i \rangle) = \varphi(X)$$

This shows that the system S' satisfies the condition (1).

We now proceed with S' in place of S and consider the linear relation

$$(4) \qquad a_1 x_1' + \sum_{i=r+1}^{k} a_i x_i = 0$$

which follows from (2). Since the number of non-zero coefficients of (4) is less than that of (2) we may repeat the procedure until we obtain the case l = 1.

We have therefore shown that there exists a system $S = \{x_1, \ldots, x_n\}$ of n generators satisfying

$$\sum_{i=1}^{n} \varphi(\langle x_i \rangle) = \varphi(X).$$

Since $\dim_{\bar{K}} \varphi(X) = n$ the vectors $x_1, \ldots x_n$ are normally independent taking account of 9.2.

EXAMPLE : Let K = k(t) be the rational function field over a commutative field k. We consider the usual t-adic valuation on K. Further, let T = K×K×K be the valued vector space over K with the cubical norm.

Now let U be the module spanned by

$$x_1 = (t,\ t+t^2,\ t^3), \qquad x_2 = (1,\ 1,\ 0)$$

$$x_3 = (0,\ t^3,\ 0), \qquad x_4 = (1,\ 1,\ t^6)$$

over the valuation ring B associated with the t-adic valuation. We will determine a normally independent system of generators of U using the method indicated in the proof of 11.1.

Obviously, the vectors x_i span a 3-dimensional space over K, hence X = T.

The first step is to get a system of generators which satisfies the condition (1). For the given system S of generators $\overline{S}$ is 2-dimensional. But the system

$$S' = \{x_1, x_2, x_3, x_4'\}$$

with

$$x_4' = x_4 - x_2 = (0, 0, t^6)$$

satisfies the condition (1).

A simple calculation gives the linear relation

$$t^3 x_1 - t^4 x_2 - t^2 x_3 - x_4' = 0.$$

The first two summands have maximal norm. Hence we must replace x_1 by

$$x_1' = x_1 - t\, x_2 = (0, t^2, t^3).$$

The linear relation shortens to

$$t^3 x_1' - t^2 x_3 - x_4' = 0$$

Again the first two summands have maximal norm. Hence we must substitute

$$x_3' = -x_3 + t x_1' = (0, 0, t^4)$$

In the next step we obtain the null-vector. Therefore, the vectors

$$x_1' = (0, t^2, t^3), \quad x_2 = (1, 1, 0), \quad x_3' = (0, 0, t^4)$$

form a normally independent system of generators of U.

We shall now apply 11.1 to the modules $U = X_1 + Y_1$ where X and Y are finite dimensional subspaces of the valued vector space T. The mapping φ of the lattice of subspaces of T onto the lattice of subspaces of the residue class space

$$\varphi(X) = \{ \bar{x} \in \bar{T} \mid x \in X_1 \}$$

preserves inclusion but is not a lattice homomorphism. In general

$$\varphi(X \cap Y) \subseteq \varphi(X) \cap \varphi(Y) \quad \text{and} \quad \varphi(X) + \varphi(Y) \subseteq \varphi(X + Y)$$

is valid. Let us denote the dimensions of the spaces occuring there

$k = \dim_{\bar{K}}\varphi(X) = \dim_K X$

$n = \dim_{\bar{K}}\varphi(Y) = \dim_K Y$

$s = \dim_{\bar{K}}\varphi(X \cap Y) = \dim_K(X \cap Y)$

$t = \dim_{\bar{K}}(\varphi(X) \cap \varphi(Y))$

By the dimension formula we obtain

$\dim_{\bar{K}}\varphi(X + Y) = \dim_K(X + Y) = k + n - s$

$\dim_{\bar{K}}(\varphi(X) + \varphi(Y)) = k + n - t.$

By 9.3 there exists a subspace Y_0 of Y such that

$$\varphi(X) + \varphi(Y) = \varphi(X) + \varphi(Y_0)$$

and a subspace Z of $X + Y$ such that

$$\varphi(X + Y) = \varphi(X) + \varphi(Y_0) + \varphi(Z)$$

where the sums on the right hand side are direct. The direct decomposition of $\varphi(X + Y)$ in the residue class space corresponds to a normally independent decomposition

$$X + Y = X + Y_0 + Z$$

in T. It is easy to compute the dimensions

$$\dim_K Y_0 = n - t \quad \text{and} \quad \dim_K Z = t - s.$$

THEOREM 11.2. Let X, Y be subspaces of the valued vector space T and k, n, s, t be defined as above. Then there exist normal bases

$$\{x_1, \ldots, x_k\} \text{ of } X \text{ and } \{y_1, \ldots, y_n\} \text{ of } Y$$

such that

(1) $x_i = y_i$ for $i = 1, \ldots, s$

$0 < \|y_i - x_i\| < |1|$ for $i = s+1, \ldots, t$.

(2) if we put $w_i = y_i - x_i$, then the vectors

$$x_1, \ldots, x_k, y_{t+1}, \ldots, y_n, w_{s+1}, \ldots, w_t$$

form a normally independent system of generators of $X_1 + Y_1$.

PROOF : Since $\varphi(X \cap Y) \subseteq \varphi(X) \cap \varphi(Y) \subseteq \varphi(X)$ we may choose a basis

$\{v_1, \ldots, v_s\}$ of $\varphi(X \cap Y)$,

and extend it to a basis

$\{v_1, \ldots, v_s, v_{s+1}, \ldots, v_t\}$ of $\varphi(X) \cap \varphi(Y)$

and then to a basis

$\{v_1, \ldots, v_s, v_{s+1}, \ldots, v_t, v_{t+1}, \ldots, v_k\}$ of $\varphi(X)$

Finally, we choose a basis

$\{u_{t+1}, \ldots, u_n\}$ of $\varphi(Y_0)$.

Since $\varphi(Y) = (\varphi(X) \cap \varphi(Y)) + \varphi(Y_0)$, the vectors

$$v_1, \ldots, v_t, u_{t+1}, \ldots, u_n$$

form a basis of $\varphi(Y)$. With respect to the residue class homomorphism, there are inverse images

a) $x_1 = y_1, \ldots, x_s = y_s$ in $X \cap Y$ of $v_1, \ldots, v_s$

b) $x'_{s+1}, \ldots, x'_t$ in X of $v_{s+1}, \ldots, v_t$

$y'_{s+1}, \ldots, y'_t$ in Y of $v_{s+1}, \ldots, v_t$

c) $x_{t+1}, \ldots, x_k$ in X of $v_{t+1}, \ldots, v_k$

$y_{t+1}, \ldots, y_n$ in Y of $u_{t+1}, \ldots, u_n$

We must replace the vectors in b) by vectors x_i and y_i such that $y_i - x_i$ are normally independent in Z, $i = s+1, \ldots, t$, where

$$(*) \qquad X + Y = X + Y_0 + Z$$

is the normal decomposition considered above.

1.) First, we have with respect to (*),

$$y_i' - x_i' = l_i + k_i + z_i, \quad l_i \in X,\ k_i \in Y_0,\ z_i \in Z,$$

for $i = s+1, \ldots, t$. Since y_i' and x_i' have the same image under the residue class homomorphism, we have $y_i' - x_i' \in T_M$ and therefore $l_i, k_i, z_i \in T_M$ because the sum of the right hand side of (*) is normally independent. Now replace x_i' by $x_i' + l_i$ and y_i' by $y_i' - k_i$. Under the residue class homomorphism the images of the new vectors are also v_i. Their difference lies in Z. Hence we may assume

$$y_i' - x_i' = z_i \in Z \cap T_M \quad \text{for } i = s+1, \ldots, t.$$

2.) The k+n-s vectors listed in a), b), and c) span X + Y. Since the dimension of X + Y is k+n-s, they are linearly independent over K. In particular,

$$x'_{s+1}, \ldots, x'_t \quad \text{and} \quad y'_{s+1}, \ldots, y'_t$$

and therefore

$$z_{s+1}, \ldots, z_t$$

are linearly independent over K. Since Z has dimension t-s the vectors

$$z_{s+1}, \ldots, z_t$$

form a basis of Z.

3.) Consider the B-module W, generated by

$$z_{s+1}, \ldots, z_t$$

which is a submodule of T_M since $z_i \in T_M$. By 11.1 W has a normally independent system of t-s generators

$$w_{s+1}, \ldots, w_t$$

hence

$$W = \langle z_{s+1}, \ldots, z_t \rangle_B = \langle w_{s+1}, \ldots, w_t \rangle_B.$$

There exist matrices (a_{ik}) and (b_{ik}) with coefficients in B such that

$$w_i = \sum_{k=s+1}^{t} a_{ik} z_k \quad \text{and} \quad z_i = \sum_{k=s+1}^{t} b_{ik} w_k$$

Both matrices are inverse to each other. Applying the residue class homomorphism $B \longrightarrow \overline{K}$, $a \longmapsto \bar{a}$, we obtain a regular matrix $(\overline{a_{ik}})$ over $\overline{K}$.

Now define

$$x_i = \sum_{k=s+1}^{t} a_{ik} x_k'$$

$$y_i = \sum_{k=s+1}^{t} a_{ik} y_k' = x_i + w_i$$

for $i = s+1, \ldots, t$. Since $(\overline{a_{ik}})$ is a regular matrix, the vectors

$$\overline{x_i} = \overline{y_i} = \sum_{j=s+1}^{t} \overline{a_{ik}} v_k, \quad i = s+1, \ldots, t$$

are linearly independent over $\overline{K}$. Change the basis of $\varphi(X) \cap \varphi(Y)$. Instead of

$$v_{s+1}, \ldots, v_t$$

choose

$$\overline{x_{s+1}}, \ldots, \overline{x_t}.$$

Then the inverse images x_i in X and y_i in Y have the property that $y_i - x_i = w_i \in Z$, $i = s+1, \ldots, t$, are normally independent.

We are now ready to check the assertions of the theorem. Since the vectors $x_1, \ldots, x_k$ and $y_1, \ldots, y_n$ are inverse images of linearly independent vectors of $\varphi(X)$ and $\varphi(Y)$ they form normal bases of X and Y. This implies

$$X_1 + Y_1 = \langle x_1, \ldots, x_k, y_1, \ldots, y_n \rangle_B$$

$$= \langle x_1, \ldots, x_k, y_{t+1}, \ldots, y_n, w_{s+1}, \ldots, w_t \rangle_B$$

By definition $x_1, \ldots, x_k \in X$, $y_{t+1}, \ldots, y_n \in Y_0$, and $w_{s+1}, \ldots, w_t \in Z$. These subspaces are normally independent, hence the vectors are normally independent.

Since $w_i \in W \subseteq T_M$ we have $\|w_i\| < |1|$ for $i = s+1, \ldots, t$.

12. The I-hull and the I-kernel of a submodule

Let T be a valued vector space over K and I a proper ideal of the valuation ring B.

As defined in section 9

$$T_I = \{ x \in T \mid \|x\| = |a| \text{ for some } a \in I \}$$

is a submodule of $T_1 = \{ x \in T \mid \|x\| \leq |1| \}$.

Let U be any submodule of T_1. Consider the set of all subspaces X of T such that

$$U \subseteq X_1 + T_I .$$

We now define the **I-hull** $h_I(U)$ as the set of all these subspaces which are minimal with this property.

THEOREM 12.1. Let

$$U = \sum_{i=1}^{n} B\, u_i$$

be a submodule of T_1 with normally independent generators u_i. Let the indices be chosen such that

$$u_i \notin T_I \quad \text{for} \quad i = 1, \ldots, r$$

$$u_i \in T_I \quad \text{for} \quad i = r+1, \ldots, n.$$

Then $h_I(U)$ consists of all subspaces

$$X = \langle u_1 + t_1, \ \ldots \ , u_r + t_r \rangle_K$$

where $t_i \in T_I$.

PROOF : 1.) Let $X \in h_I(U)$. Since $U \subseteq X_1 + T_I$ we have

$$u_i = x_i - t_i , \quad x_i \in X_1 , \quad t_i \in T_I$$

for $i = 1, \ldots, r$. Let X′ be the subspace of X generated by

$$x_1, \ldots , x_r .$$

Then $X' = X$ because $U \subseteq X'_1 + T_I$ and X is minimal with this property. Hence

$$X = \langle u_1 + t_1, \ldots, u_r + t_r \rangle_K .$$

2.) Conversely, let

$$X = \langle u_1 + t_1, \ldots, u_r + t_r \rangle_K$$

with $t_i \in T_I$. Clearly $U \subseteq X_1 + T_I$. We must show the minimality of X. Let Y be a subspace of X such that $U \subseteq Y_1 + T_I$. Then $u_i = y_i + t_i$ where $y_i \in Y_1$ and $t_i \in T_I$, $i = 1, \ldots, r$. Since the u_i are normally independent and $u_i \notin T_I$ the vectors y_i, $i = 1, \ldots, r$, are also normally independent, hence linearly independent, thus $\dim Y \geq r$. On the other hand, we have $\dim X \leq r$, hence $X = Y$.

Since the definition of the I-hull does not depend on the choice of the generators of U, the number r is an invariant of U depending only on the ideal I. We call it the **I-rank** of U, denoted by $rg_I(U)$. In 12.1 $\dim X = rg_I(U)$ was proved. If U_1 and U_2 are normally independent then we have

$$rg_I(U_1 \dot{+} U_2) = rg_I(U_1) + rg_I(U_2).$$

Consider the special case $I = M$. Applying the residue class homomorphism $x \longmapsto \bar{x}$ onto U we obtain a subspace

$$\bar{U} = \{ \bar{u} \mid u \in U \}$$

of the residue class space. Its dimension is the M-rank, hence

$$rg_M(U) = \dim_{\bar{K}} \bar{U}.$$

The set of all subspaces X of T satisfying

$$X_1 \subseteq U + T_I$$

and being maximal under this condition, is called the **I-kernel** $k_I(U)$. By Zorn's lemma $k_I(U)$ is not empty.

PROPOSITION 12.2. We have $\varphi(X) = \overline{U}$ for any $X \in k_I(U)$.

PROOF : From $X_1 \subseteq U + T_I$ we obtain $\varphi(X) \subseteq \overline{U}$. Suppose that there exists an element $u \in U$ such that $\bar{u} \in \overline{U}$, but $\bar{u} \notin \varphi(X)$. Put $X' = X + \langle u\rangle$. Since X and $\langle u\rangle$ are normally independent, we have $X'_1 = X_1 + Bu$ and therefore $X'_1 \subseteq U + T_I$, a contradiction to the maximality of X.

COROLLARY 12.3. We have $\dim X = rg_M U$ for any $X \in k_I(U)$.

PROOF : Since T is a valued vector space, by 10.1

$$\dim_K X = \dim_{\overline{K}} \varphi(X) = \dim_{\overline{K}} \overline{U} = rg_M U.$$

If $rg_M U$ is finite, the maximality condition may be replaced by $\dim X = rg_M U$, hence

$$k_I(U) = \{ X \mid X_1 \subseteq U + T_I,\ \dim X = rg_M U \}.$$

A submodule U' of a module U is called a **remainder** of U when $U' \subseteq T_M$ and when there exists a subspace X of T which is normally independent to U' such that $U = X_1 + U'$.

Since

$$rg_M U = rg_M X_1 + rg_M U' = \dim X$$

the subspace X is an element of $k_I(U)$.

If U is finitely generated, say

$$U = \sum_{i=1}^{n} Bu_i$$

where the vectors u_i are normally independent and

$\|u_i\| = |1|$ for $i = 1, \dots, k-1$

$\|u_i\| < |1|$ for $i = k, \dots, n$

then

$$U' = \sum_{i=k}^{n} Bu_i$$

is a remainder of U. If we only assume that U spans a finite dimensional subspace of T, then the proof of the existence of U′ is a little more difficult.

PROPOSITION 12.4. If the subspace X of T spanned by U is finite dimensional then U possesses a remainder.

PROOF : Let Y be a subspace of X such that

$$\varphi(X) = \varphi(Y) + \overline{U}, \qquad \varphi(Y) \cap \overline{U} = 0.$$

Put $U' = Y \cap U$.

Let $u \in U$, $\|u\| = |1|$. Then $\bar{u} \in \overline{U}$, hence $\bar{u} \notin \varphi(Y)$, hence $\bar{u} \notin U'$. From this it follows that $u \in U'$ implies $\|u\| < |1|$, hence $U' \subseteq T_M$.

Let Z′ be an arbitrary element of $k_I(U)$. Since $\dim Z' \leq \dim X < \infty$ there exists a normal basis $\{z_1, \dots, z_k\}$ of Z′. From $Z'_1 \subseteq U + T_I$ it follows $z_i = u_i + t_i$ with $u_i \in U$, $t_i \in T_I$. We define $Z = \langle u_1, \dots, u_k \rangle$. Then $\{u_1, \dots, u_k\}$ is a normal basis of Z. Hence $Z \in k_I(U)$ and $Z_1 \subseteq U$. Since $\varphi(Z) = \overline{U}$ we have $\varphi(Z) \cap \varphi(Y) = 0$. Therefore Z and Y are normally independent.

Further, $Z_1 \subseteq U$ implies $Z \subseteq X$, hence $Z + Y \subseteq X$. Since

$$\begin{aligned} \dim_K(Z + Y) &= \dim_K Z + \dim_K Y \\ &= \dim_{\overline{K}}\overline{U} + \dim_{\overline{K}}\varphi(Y) \\ &= \dim_{\overline{K}}\varphi(X) = \dim_K X \end{aligned}$$

we have $Z + Y = X$.

Now we are ready to show $U = Z_1 + U'$ from which follows that U′ is a remainder of U. Let $u \in U \subseteq X$ and $u = z + y$ with $z \in Z$, $y \in Y$. Since Z and Y are normally independent we have $\|u\| = \mathrm{Max}(\|z\|, \|y\|)$ which implies $\|z\| \leq \|u\| \leq |1|$, hence $z \in Z_1$. Moreover, $Z_1 \subseteq U$ implies $z \in U$, hence $y = u - z \in Y \cap U = U'$. Thus $u = z + y \in Z_1 + U'$. The inverse inclusion is clear since Z_1 and U′ are both contained in U.

THEOREM 12.5. Let U be a submodule of T_1 such that the subspace spanned by U is finite dimensional. If U' is a remainder of U and X an arbitrary element of $k_I(U)$ with the normal basis $\{x_1, \ldots, x_k\}$ then

$$k_I(U) = \{ Z = \langle x_1 + u_1, \ldots, x_k + u_k \rangle \mid u_i \in U' + T_I \}$$

PROOF : Let $Z \in k_I(U)$. First we show

$$U + T_I = Z_1 + U' + T_I.$$

Let $\{z_1, \ldots, z_k\}$ be a normal basis of Z. Since $Z_1 \subseteq U + T_I$ we have

$$z_i = v_i + t_i, \quad v_i \in U, \quad t \in T_I, \quad i = 1, \ldots, k.$$

Put $Y = \langle v_1, \ldots, v_k \rangle$. Since z_i and v_i have the same images under the residue class homomorphism the vectors v_i form a normal basis of Y. Further, $Y \in k_I(U)$ because $\dim Y = k = rg_M U$ and $Y_1 = \Sigma Bv_i \subseteq U + T_I$. As was shown in the proof of 12.4 $U = Y_1 + U'$, hence

$$U + T_I = Y_1 + U' + T_I = Z_1 + U' + T_I.$$

Now, $X_1 \subseteq U + T_I = Z_1 + U' + T_I$ implies

$$x_i = z_i - u_i, \quad z_i \in Z_1, \quad u_i \in U' + T_I, \; i = 1, \ldots, k.$$

Then $\{z_1, \ldots, z_k\}$ is a normal basis of Z and

$$Z = \langle x_1 + u_1, \ldots, x_k + u_k \rangle.$$

Conversely, a subspace $Z = \langle x_1 + u_1, \ldots, x_k + u_k \rangle$ with $u_i \in U' + T_I$ satisfies

$$Z_1 = \Sigma B(x_i + u_i) \subseteq X_1 + U' + T_I = U + T_I$$

and $\dim Z = rg_M U$, hence $Z \in k_I(U)$.

EXERCISES

1.) Let T be a normed vector space, U a subspace of T, and $x \in T$, $x \notin U$. Assume that there exists a best approximation $x_0 \in U$ of x. Show that $x-x_0$ and U are normally independent.

2.) Let T be a normed vector space. Show that T is a valued vector space if and only if for every finite dimensional subspace U of T and any subspace X of U there exists a subspace Y such that $U = X + Y$ and X and Y are normally independent.

3.) Let K be a field with a discrete valuation and I a denumerable index set. Let T be the subspace of the direct product K^I consisting of all vectors $x = (a_i)$, $i \in I$, for which

$$I_\varepsilon = \{ i \in I \mid |a_i| > \varepsilon \}$$

is finite for all $\varepsilon \in W^\times$. Define

$$\|x\| = \operatorname{Max}(|a_i|, \ i \in I).$$

Show that T becomes a valued vector space without a normal basis.

4.) Let $K = k(t)$ be the rational function field over a commutative field k with the t-adic valuation. Let T be the valued vector space of the 5-tuples over K with the cubical norm. Let X be the subspace spanned by the vectors

$$(1,0,0,0,0), \quad (0,1,0,0,0), \quad (0,0,1,0,0)$$

and Y the subspace spanned by the vectors

$$(1,t,0,t^2,0), \quad (0,1+t,t,t^2,t^3)$$

Proceed as in the proof of 11.2 to get a normally independent system of generators of $X_1 + Y_1$.

5.) Let T be a valued vector space and U and V two finitely generated normally independent B-submodules of T_1. Show that the I-hull of $U + V$ consists of all subspaces $X + Y$ with $X \in h_I(U)$ and $Y \in h_I(U)$.

CHAPTER 4 PROJECTIVE HJELMSLEV SPACES

Let F be a free module over a local ring. The set H(F) of all direct summands of F is called the projective Hjelmslev space belonging to F. Characteristic for these geometries is the fact that H(F) is generally not a lattice, it is only partially ordered by inclusion. This means geometrically that join and intersection are, in general, not unique.

We shall investigate special kinds of projective Hjelmslev spaces which are induced by valued vector spaces. The main result is a description of the union and intersection sets. This implies the dimension formula in these geometries.

Throughout this chapter T denotes a valued vector space of finite dimension over K and I a proper two-sided ideal of the valuation ring B.

13. Definition of the projective Hjelmslev spaces H_I.

Let T be a valued vector space over K of dimension n and I a proper two-sided ideal of B.

We define H_I as the set of all direct summands of the factor module

$$F = T_1/T_I .$$

Let $f_I : T_1 \longrightarrow F$ denote the canonical homomorphism. There exists a mapping φ_I from the set of all subspaces X of T to the set of all submodules of F defined by

$$\varphi_I(X) = \{ f_I(x) \mid x \in X_1 = X \cap T_1 \}.$$

An important special case is I = M. Then, f_M is the residue class homomorphism and φ_M is the mapping φ considered in section 9.

THEOREM 13.1. For a submodule V of F the following assertions are equivalent

(1) V is a direct summand of F.

(2) There exists a subspace X of T such that $V = \varphi_I(X)$.

PROOF : (1) ⇒ (2). Since T has finite dimension, F is a finitely generated B-module. Let V be a direct summand of F. Then V is also finitely generated because it is a homomorphic image of F. Let $\{v_1, \ldots, v_k\}$ be a system of generators of V and $\{u_1, \ldots, u_k\}$ elements of T_1 such that $f_I(u_i) = v_i$. If

$$U = \langle u_1, \ldots, u_k \rangle_B$$

is the B-submodule of T_1 spanned by the vectors u_i then $f_I(U) = V$. By 11.1 U has a system of normally independent generators. Hence we may assume that the system $\{u_1, \ldots, u_k\}$ is already normally independent and that $f_I(u_i) = v_i \neq 0$ for $i = 1, \ldots, k$.

We will show $\|u_i\| = |1|$ for all i. Suppose $\|u_i\| < |1|$ for some i. Normalizing u_i we obtain

$$u_i = a\, u_i' \quad \text{with } a \in M,\ \|u_i'\| = |1|.$$

Since V is a direct summand of F, we have a direct decomposition

$$F = V + V',$$

hence

$$f_I(u_i') = x_i + x_i' \quad \text{with } x_i \in V,\ x_i' \in V'.$$

Since

$$f_I(u_i) = a\, f_I(u_i') = ax_i + ax_i' \in V$$

we get $ax_i' = 0$, hence $f_I(u_i) = ax_i$.

Let

$$x_i = \sum_{j=1}^{k} b_j v_j = f_I\left(\sum_{j=1}^{k} b_j u_j\right), \quad b_j \in B.$$

Then

$$f_I(u_i) = a\, x_i = f_I(a \sum_{j=1}^{k} b_j u_j),$$

hence

$$u_i - a \sum_{j=1}^{k} b_j u_j \in T_I$$

implies, since $u_1, \dots, u_k$ are normally independent

$$(1 - a\, b_i) u_i \in T_I .$$

From $a \in M$ follows that $1 - a\, b_i$ is invertible in B, hence $u_i \in T_I$ contrary to $f_I(u_i) = v_i \neq 0$

Thus we have shown $\|u_i\| = |1|$ for $i = 1, \dots, k$. Let

$$X = \langle u_1, \dots, u_k \rangle$$

be the subspace spanned by the vectors u_i. Then $\{u_1, \dots, u_k\}$ is a normal basis of X and

$$X_1 = X \cap T_1 = \langle u_1, \dots, u_k \rangle_B = U,$$

hence

$$\varphi_I(X) = f_I(X_1) = \langle v_1, \dots, v_k \rangle = V.$$

(2) ⇒ (1). Let $V = \varphi_I(X)$ where X is a subspace of T. In the residue class space there exists a complement of $\varphi(X)$. Let Y be a subspace of T which is an inverse image of this complement. Then X and Y are normally independent. Now it is easily shown that

$$F = \varphi_I(X) + \varphi_I(Y), \quad \varphi_I(X) \cap \varphi_I(Y) = 0.$$

Hence, $V = \varphi_I(X)$ is a direct summand in F.

By 13.1 we have

$$H_I = \{ \varphi_I(X) \mid X \text{ subspace of } T \}.$$

Therefore φ_I is a surjective mapping from the lattice of subspaces of T onto H_I, but it is not injective:
$\varphi_I(X) = \varphi_I(X')$ holds if and only if $X_1 + T_I = X'_1 + T_I$.

PROPOSITION 13.2. Let $\{x_1, \ldots, x_k\}$ be a normal basis of X.

(1) If $\varphi_I(X) = \varphi_I(X')$ then there exist vectors $t_i \in T_I$ such that

$$X' = \langle x_1 + t_1, \ldots, x_k + t_k \rangle$$

(2) If $X' = \langle x_1 + t_1, \ldots, x_k + t_k \rangle$, $t_i \in T_I$, then $\varphi_I(X) = \varphi_I(X')$.

(3) Let $X' = \langle x_1 + u, x_2, \ldots, x_k \rangle$ where u is normally independent to X and not contained in T_I. Then $\varphi_I(X) \neq \varphi_I(X')$.

PROOF : The proof of (1) and (2) is straightforward and therefore omitted. (3) Let us assume $\varphi_I(X) = \varphi_I(X')$. Then $X_1 + T_I = X'_1 + T_I$. Since $x_1 + u \in X_1 + T_I$ there exists $x \in X_1$ such that $x + u = t \in T_I$. We obtain

$$\|t\| = \|x + u\| = \mathrm{Max}(\|x\|, \|u\|)$$

because x and u are normally independent, hence $\|u\| \leq \|t\|$. This implies $u \in T_I$, a contradiction.

PROPOSITION 13.3. Let $\varphi_I(X) \subseteq V \subseteq \varphi_I(X')$ in H_I and $X \subseteq X'$. Then there exists a subspace Y of T such that $X \subseteq Y \subseteq X'$ and $\varphi_I(Y) = V$.

PROOF : First let Y' be an arbitrary subspace of T such that $\varphi_I(Y') = V$. Since $\varphi_I(X) \subseteq V$ we have $X_1 \subseteq Y'_1 + T_I$. Hence, if $\{x_1, \ldots, x_k\}$ is a normal basis of X there exist $t_i \in T_I$ and $y_i \in Y'_1$ such that

$$x_i = y_i + t_i, \quad i = 1, \ldots, k.$$

Let

$$\{y_1, \ldots, y_k, y_{k+1}, \ldots, y_s\}$$

be a normal basis of of Y'. Since $Y'_1 \subseteq X'_1 + T_I$ we have

$$y_i = x_i + t_i, \quad x_i \in X'_1, \quad t_i \in T_I, \quad i = k+1, \ldots, s.$$

Then

$$Y = \langle x_1, \ldots, x_k, x_{k+1}, \ldots, x_s \rangle_K$$

satisfies the conditions desired.

In order to simplify notation we write

$$\tilde{X} = \varphi_I(X)$$

where X is a subspace of T. We define

$$\dim \tilde{X} = \dim_K X.$$

By 13.2 this definition is in fact independent of the choice of X. Moreover, the dimension function on H_I has the customary properties of a dimension function of a finite dimensional vector space. Elements of dimension 1 are called points, elements of dimension 2 are called lines etc. H_I is said to be a projective Hjelmslev plane if it has dimension 3.

Further, there is a mapping φ' from H_I onto the lattice of subspaces of the residue class space defined by $\varphi_I(X) \longmapsto \varphi(X)$. This mapping preserves the inclusion and the dimension of the elements of H_I. Geometric structures of this kind were first investigated by Klingenberg [3].

14. The union and intersection sets in H_I

The projective Hjelmslev spaces are generally not lattices, but partially ordered sets. Instead of union and intersection we introduce the notions of union and intersection sets. The union set $V(\tilde{X}, \tilde{Y})$ of two elements $\tilde{X}$ and $\tilde{Y}$ of H_I is defined as the set of all minimal upper bounds and the intersection set $S(\tilde{X}, \tilde{Y})$ as the set of all maximal lower bounds of $\tilde{X}$ and $\tilde{Y}$.

PROPOSITION 14.1. Let $\tilde{X} = \varphi_I(X)$, $\tilde{Y} = \varphi_I(Y)$ be elements of H_I. Then

$$V(\tilde{X}, \tilde{Y}) = \{ \tilde{Z} = \varphi_I(Z) \in H_I \mid Z \in h_I(V) \}$$

$$S(\tilde{X}, \tilde{Y}) = \{ \tilde{Z} = \varphi_I(Z) \in H_I \mid Z \in k_I(W) \}$$

where $V = X_1 + Y_1$ and $W = (X_1 + T_I) \cap (Y_1 + T_I)$.

PROOF : $\tilde{X}, \tilde{Y} \leq \tilde{Z}$ is equivalent to $X_1, Y_1 \subseteq Z_1 + T_I$ or to $V = X_1 + Y_1 \subseteq Z_1 + T_I$. Therefore the formula for the union set follows immediately from the definition of the I-hull. The the formula for the intersection set is similarly obtained.

In order to determine the union set of two elements, we use 11.2. By this theorem there exist normal bases $\{x_1, \ldots, x_k\}$ of X and $\{y_1, \ldots, y_n\}$ of Y such that

$$x_1, \ldots, x_k, y_{t+1}, \ldots, y_n, w_{s+1}, \ldots, w_t$$

are normally independent generators of $X_1 + Y_1$ where $w_i = y_i - x_i$ for $i = 1, \ldots, t$ and

$$s = \dim_K(X \cap Y) \text{ and } t = \dim_{\bar{K}}(\varphi(X) \cap \varphi(Y)).$$

For $i = 1, \ldots, s$ we have $w_i = 0$, hence $x_i = y_i$ and for $i = s+1, \ldots, t$

$$0 < \|w_i\| < |1|.$$

We may further assume

$$\|w_1\| \leq \ldots \leq \|w_t\|.$$

With respect to the ideal I, we define the number r by

$$w_1, \dots, w_r \in T_I$$

$$w_{r+1}, \dots, w_t \notin T_I.$$

Applying 12.1 we then obtain

THEOREM 14.2. Let $\tilde{X}, \tilde{Y} \in H_I$ and k, n, r, and t be defined as above. Then the elements of the union set have the dimension k + n - r and can be described as $\varphi_I(Z)$ with

$$Z = \langle x_1+u_1, \dots, x_k+u_k, y_{t+1}+u'_{t+1}, \dots, y_n+u'_n, w_{r+1}+v_{r+1}, \dots, w_t+v_t \rangle$$

where u_i, u_i', $v_i \in T_I$.

We shall now determine the intersection sets of two elements of H_I. Let $w \in T_1$. Then $\{b \in B \mid bw \in T_I\}$ is a left ideal of B. If $\|w\| = |a|$, $a \in B$, this ideal is equal to $Ia^{-1} = \{b \in B \mid ba \in I\}$ and depends therefore only on the norm $\|w\|$. Hence we write

$$I\|w\|^{-1} = \{b \in B \mid bw \in T_I\}.$$

It is easily shown that $w \in T_I$ if and only if $I\|w\|^{-1} = B$. It should be noted that $I\|w\|^{-1}$ is also defined if w = 0.

In the following we shall use the normal bases $\{x_1, \dots, x_k\}$ of X and $\{y_1, \dots, y_n\}$ of Y constructed in 11.2. For $w_i = y_i - x_i$, $i = 1, \dots, t$, we put

$$I_i = I\|w_i\|^{-1}$$

and

$$U = \sum_{i=1}^{t} I_i x_i$$

PROPOSITION 14.3. We have

$$(X_1 + T_I) \cap (Y_1 + T_I) = U + T_I.$$

PROOF : 1.) First we show $X_1 \cap (Y_1 + T_I) \subseteq U + T_I$. Let $x \in X_1 \cap (Y_1 + T_I)$, hence

$$x = \sum_{i=1}^{k} b_i x_i, \quad b_i \in B$$

and $x = y + v, \quad y \in Y_1, v \in T_I$.

Expressing y by the basis vectors of Y we obtain

$$y = \sum_{i=1}^{n} c_i y_i, \quad c_i \in B$$

and putting $y_i = x_i + w_i$ for $i = 1, \ldots, t$

$$v = x - y = \sum_{i=1}^{t} (b_i - c_i)x_i + \sum_{i=t+1}^{k} b_i x_i - \sum_{i=1}^{t} c_i w_i - \sum_{i=t+1}^{n} c_i y_i$$

Since $v \in T_I$, the normal independence yields

$$(b_i - c_i)x_i \in T_I \quad \text{and} \quad c_i w_i \in T_I$$

for $i = 1, \ldots, t$ and

$$\sum_{i=t+1}^{k} b_i x_i \in T_I.$$

Since $\|x_i\| = |1|$ we obtain $b_i - c_i \in I$, hence $b_i w_i = (b_i - c_i)w_i + c_i w_i \in T_I$, hence $b_i \in I_i$ for $i = 1, \ldots, t$. Therefore

$$x = \sum_{i=1}^{k} b_i x_i = \sum_{i=1}^{t} b_i x_i + \sum_{i=t+1}^{k} b_i x_i \in U + T_I.$$

2.) Now we show $(X_1 + T_I) \cap (Y_1 + T_I) \subseteq U + T_I$. Let $v \in (X_1 + T_I) \cap (Y_1 + T_I)$. Then $v = x + u = y + u'$ with $x \in X_1$, $y \in Y_1$, $u, u' \in T_I$. We obtain

$$x = y + u - u' \in X_1 \cap (Y_1 + T_I) \subseteq U + T_I$$

hence

$$v = x + u \in U + T_I.$$

3.) By the definition of I_i we have $I_i w_i \subseteq T_I$, hence

$$U + T_I \subseteq \sum_{i=1}^{t} I_i x_i + T_I = \sum_{i=1}^{t} I_i(y_i - w_i) + T_I$$

$$\subseteq \sum_{i=1}^{t} I_i y_i + T_I \subseteq Y_1 + T_I$$

and obviously

$$U + T_I \subseteq X_1 + T_I,$$

hence

$$U + T_I \subseteq (X_1 + T_I) \cap (Y_1 + T_I).$$

We are now ready to determine the I-kernel of

$$W = (X_1 + T_I) \cap (Y_1 + T_I).$$

Since $w_i \in T_I$ is equivalent to $I_i = B$ we have $I_i = B$ for $i = 1, \ldots, r$ and $I_i \subseteq M$ for $i = r+1, \ldots, t$ where M is the maximal ideal of B. Thus

$$U = \sum_{i=1}^{r} Bx_i + \sum_{i=r+1}^{t} I_i x_i = \sum_{i=1}^{r} Bx_i + U'$$

where

$$U' = \sum_{i=r+1}^{t} I_i x_i \subseteq T_M$$

is a remainder of U.

Now applying 12.5 we obtain

THEOREM 14.5. Let $\tilde{X}, \tilde{Y} \in H_I$. Then the elements of the intersection set $S(\tilde{X}, \tilde{Y})$ have the dimension r and can be described as $\tilde{Z} = \varphi_I(Z)$ where

$$Z = \langle x_1 + u_1, \ldots, x_r + u_r \rangle_K$$

and $u_i \in U' + T_I$.

Since the dimension of the elements of the union set $S(\tilde{X}, \tilde{Y})$ is $k + n - r$, where $k = \dim \tilde{X}$, $n = \dim \tilde{Y}$, we obtain

THEOREM 14.6. In the projective Hjelmslev spaces H_I the dimension formula

$$\dim \tilde{X} + \dim \tilde{Y} = \dim V(\tilde{X}, \tilde{Y}) + \dim S(\tilde{X}, \tilde{Y})$$

is valid.

REMARK : The validity of the dimension formula is characteris for the spaces H_I. It does not hold, in general, in projective Hjelmslev spaces over local rings (see exercise 1.).

The projective Hjelmslev spaces H_I for distinct ideals are closely related to each other. Let I, I′ be two ideals of B. If $I' \subseteq I$ we define a mapping

$$\varphi_{I,I'} : H_{I'} \longrightarrow H_I, \quad \varphi_{I'}(X) \longmapsto \varphi_I(X)$$

which preserves inclusion and dimension of the elements. According to the linear order of the ideal chain of B

$$O \subseteq \ldots \subseteq I' \subseteq I \subseteq \ldots \subseteq M$$

we obtain a linear inductive system of projective Hjelmslev spaces

$$H_O \longrightarrow \ldots \longrightarrow H_{I'} \longrightarrow H_I \longrightarrow \ldots \longrightarrow H_M.$$

Immediately from 14.3 and 14.5 follows that $\varphi_{M,I}$ maps all elements of $V(\tilde{X}, \tilde{Y})$ resp. $S(\tilde{X}, \tilde{Y})$ onto one element in H_M. This property is one of the conditions which Hjelmslev [1] claimed for the natural geometries. It was taken up by Klingenberg [3] in the definition of projective spaces with homomorphism (see also Machala [1]).

15. Some further properties of valuation rings

Let B be a valuation ring of K and I a proper two-sided ideal of B. We define

$$R(I) = \{ a \in K \mid Ia \subseteq I \} \quad \text{and} \quad L(I) = \{ a \in K \mid aI \subseteq I \}.$$

Obviously, R(I) and L(I) are subrings of K containing B, hence valuation rings. Applying the star-mapping we obtain completely prime ideals (conf. 1.6)

$$P_r(I) = R(I)^* \quad \text{and} \quad P_l(I) = L(I)^*$$

PROPOSITION 15.1. I is contained in $P_r(I)$ and $P_l(I)$.

PROOF : $I \subseteq P_r(I)$ is equivalent to $R(I) \subseteq I^*$. In order to show the second inclusion let $a \in R(I)$, hence $Ia \subseteq I$. Since I is a proper ideal we have $1 \notin I$, hence $1 \notin Ia$, hence $a \in I^*$. Similarly, $I \subseteq P_l(I)$ is shown.

PROPOSITION 15.2. The following assertions are equivalent

(1) I is a completely prime ideal.

(2) $I = P_r(I)$.

(3) $I = P_l(I)$.

PROOF : Since $P_r(I)$ and $P_l(I)$ are completely prime ideals (2) and (3) imply (1). Assume (1). By 15.1 $I \subseteq P_r(I)$. Therefore we must show $P_r(I) \subseteq I$ or equivalently $I^* \subseteq R(I)$. Suppose there exists $a \in I^*$, but $a \notin R(I)$. From $a \notin R(I)$ it follows that Ia is not contained in I. Hence, there exists $b \in I$ such that $ba \notin I$. From $b = (ba)a^{-1} \in I$ then follows $a^{-1} \in I$ because I is completely prime. However, this contradicts $a \in I^*$.
Similarly (1) implies (3).

An element a of a ring is said to be **right regular** if a is not a right zero-divisor. A ring with 1 is called a **right quotient ring** if each right regular element is invertible. Left regular elements and left quotient rings are similarly defined.

THEOREM 15.3. Let I be a proper two-sided ideal of the valuation ring B. Then the factor ring B/I is a right quotient ring if and only if $P_r(I) = M$. The factor ring B/I is a left quotient ring if and only if $P_l(I) = M$.

PROOF : 1.) First assume that B/I is a right quotient ring. Let $a \in R(I)$, but $a \notin B$. We show that this assumption leads to a contradiction. Since B is a valuation ring, we have $a^{-1} \in B$. We show that $a^{-1} + I$ is not a right zero-divisor in B/I. Suppose $(b + I)(a^{-1} + I) = I$. This implies $ba^{-1} \in I$, hence $b \in Ia \subseteq I$. Therefore $a^{-1} + I$ is not a right zero-divisor, henc invertible in B/I because B/I is a right quotient ring. Hence, there exists $c \in B$ such that $(c + I)(a^{-1} + I) = 1 + I$. This implies $ca^{-1} - 1 \in I$, hence $c - a \in Ia \subseteq I \subseteq B$, hence $a \in B$, a contradiction. Thus $R(I) \subseteq B$, hence $R(I) = B$. Applying the star-mapping we obtain $P_r(I) = M$ by 1.5.

2.) Assume $P_r(I) = M$ or equivalently $R(I) = B$. Let $a + I \in B/I$, $a \notin I$, be not invertible in B/I. Then a is not invertible in B. It follows $a^{-1} \notin B = R(I)$. Therefore the fractional left ideal Ia^{-1} is not contained in I. This means $I \subseteq Ia^{-1}$, $I \neq Ia^{-1}$, because the set of the fractional left ideals is totally ordered by inclusion. Hence there exists $b \in B \cap Ia^{-1}$, $b \notin I$. Then we have $(b + I)(a + I) = I$, hence $a + I$ is a right zero-divisor. This shows that B/I is a right quotient ring.

The proof of the second assertion is similar.

EXAMPLE : Let $\mathbb{Q}$ be the field of rational numbers. Then $\mathbb{Q} \times \mathbb{Q}$ becomes a totally ordered abelian group with respect to the componentwise addition and lexicographic order. By 4.1 there exists a commutative field K with an exponential valuation w whose value group is $\mathbb{Q} \times \mathbb{Q}$. The completely prime ideals of the associated valuation ring are the maximal ideal M,

$$P = \{ a \in K \mid w(a) = (r, s) > (0, 0) \text{ and } r > 0\},$$

and the null ideal.

If we embed $\mathbb{Q} \times \mathbb{Q}$ into $\mathbb{R} \times \mathbb{R}$ and if $(r, s) \in \mathbb{R} \times \mathbb{R}$, $r > 0$, then

$$I = \{ a \in K \mid w(a) > (r, s) \}$$

is an ideal lying between P and the null ideal. It is easy to show that $P_l(I) = M$ if r is rational and $P_l(I) = P$ if r is irrational.

16. Uniqueness in H_I

Two elements $\tilde{X}$, $\tilde{Y} \in H_I$ are said to have a unique join or a unique intersection when $V(\tilde{X}, \tilde{Y})$ resp. $S(\tilde{X}, \tilde{Y})$ contains only one element. If this is the case for all pairs of elements of H_I then H_I is a lattice. The dimension formula implies the modularity of the lattice. Therefore H_I is a projective space in the usual sense. We shall show that H_I is a ordinary projective space if and only if I is completely prime.

PROPOSITION 16.1. Let $w \in T_1$ and $a \in K^x$ such that $\|w\| = |a|$. Then

1.) $$w \notin T_{P_l(I)} \iff a^{-1}I \subseteq I$$

2.) $$w \notin T_{P_r(I)} \iff Ia^{-1} \subseteq I$$

PROOF : 1.) Since

$$a^{-1}I \subseteq I \iff a^{-1} \in L(I) \iff a \notin P_l(I)$$

the assertion follows from the fact that $a \in P_l(I)$ is equivalent to $w \in T_{P_l(I)}$.

2.) We have

$$Ia^{-1} \subseteq I \iff a^{-1} \in R(I) \iff a \notin P_r(I)$$

from which the assertion follows.

THEOREM 16.2. 1.) Let the join of $\tilde{X}$, $\tilde{Y} \in H_I$ be different from the whole space. Then $\tilde{X}$ and $\tilde{Y}$ have a unique join if and only if

$$\dim V(\tilde{X}, \tilde{Y}) = \dim V(\varphi_{P_l(I),I}(\tilde{X}), \varphi_{P_l(I),I}(\tilde{Y}))$$

2.) Let the intersection of $\tilde{X}$, $\tilde{Y} \in H_I$ be different from the null space. Then $\tilde{X}$ and $\tilde{Y}$ have a unique intersection if and only if

$$\dim S(\tilde{X}, \tilde{Y}) = \dim S(\varphi_{P_r(I),I}(\tilde{X}), \varphi_{P_r(I),I}(\tilde{Y}))$$

PROOF : 1.) Let $\tilde{X} = \varphi_I(X)$, $\tilde{Y} = \varphi_I(Y)$ where X and Y are subspaces of T. By 11.2 we determine normal bases

$$\{x_1, \ldots, x_k\} \text{ of } X \text{ and } \{y_1, \ldots, y_n\} \text{ of } Y$$

such that

$$x_1, \ldots, x_k, y_{t+1}, \ldots, y_n, w_{s+1}, \ldots, w_t$$

are normally independent generators of $X_1 + Y_1$. Further, we may assume that the norms of the vectors w_i increase with the indices. Normalizing these vectors we obtain

$$w_i = a_i w_i' \quad \text{with} \quad \|w_i'\| = |1|.$$

By 14.3

$$\dim V(\varphi_I(X), \varphi_I(Y)) = k + n - r$$

and

$$\dim V(\varphi_{P_1(I)}(X), \varphi_{P_1(I)}(Y)) = k + n - r',$$

where r is defined by

$$w_r \in T_I, \quad w_{r+1} \notin T_I$$

and r′ is defined by

$$w_{r'} \in T_{P_1(I)}, \quad w_{r'+1} \notin T_{P_1(I)}$$

Since $I \subseteq P_1(I)$ we have $r \leq r'$. By 16.1 $r = r'$ is equivalent to

$$a_i^{-1} I \subseteq I \quad \text{for} \quad i = r+1, \ldots, s.$$

Let

$$Z_0 = \langle x_1, \ldots, x_k, y_{t+1}, \ldots, y_n, w'_{r+1}, \ldots, w'_t \rangle$$

Then by 14.3 $\varphi_I(Z_0) \in V(\tilde{X}, \tilde{Y})$. Further by 14.3, the general element of $V(\tilde{X}, \tilde{Y})$ is $\varphi_I(Z)$ where

$$Z = \langle x_1+u_1, \ldots, x_k+u_k, y_{t+1}+u'_{t+1}, \ldots, y_n+u'_n, w_{r+1}+v_{r+1}, \ldots, w_t+v_t \rangle$$

$$= \langle x_1+u_1, \ldots, x_k+u_k, y_{t+1}+u'_{t+1}, \ldots, y_n+u'_n, w'_{r+1}+v'_{r+1}, \ldots, w'_t+v'_t \rangle$$

where u_i, u_i', $v_i \in T_I$ and $v_i' = a_i^{-1} v_i$.

If we assume $r = r'$, then $v_i' = a_i^{-1} v_i \in T_I$, hence $\varphi_I(Z) = \varphi_I(Z_0)$ by 13.2. Therefore $V(\tilde{X}, \tilde{Y})$ contains only one element.

Conversely, assume $r < r'$. Then

$$w_{r+1} \in T_{P_1(I)}, \text{ hence } a^{-1}_{r+1} I \not\subseteq I$$

Let $b \in I$ such that $a^{-1}_{r+1} b \notin I$.

Since by hypothesis $Z_0 \neq T$, there exists a vector $v \in T$, $\|v\| = |b|$, which is normal to Z_0. Then $v \in T_I$, but

$$u = a^{-1}_{r+1} v \notin T_I$$

Consider

$$Z = \langle x_1, \ldots, y_n, w_{r+1} + v, w_{r+2}, \ldots, w_t \rangle$$

$$= \langle x_1, \ldots, y_n, w'_{r+1} + u, w'_{r+2}, \ldots, w'_t \rangle$$

By 14.2 we have $\varphi_I(Z) \in V(\tilde{X}, \tilde{Y})$, by 13.2 $\varphi_I(Z) \neq \varphi_I(Z_0)$. Hence there exist two distinct elements in $V(\tilde{X}, \tilde{Y})$.

2.) We define r'' analogously to r' by

$$w_{r''} \in T_{P_1(I)}, \; w_{r''+1} \notin T_{P_r(I)}$$

The condition in the second part of the theorem is then equivalent to $r = r''$.

Assume first $r = r''$. Then

$$w_i \notin T_{P_r(I)} \quad \text{for } i = r+1, \ldots, t$$

hence $I_i = Ia_i^{-1} \subseteq I$ and therefore the remainder

$$U' = \sum_{i=r+1}^{t} I_i x_i \subseteq T_I .$$

The representation of the elements of $S(\tilde{X}, \tilde{Y})$ in 14.5 shows that the intersection is unique.

On the other hand, let $r < r''$. Then $I_{r+1} = Ia_{r+1}^{-1} \nsubseteq I$. Hence there exists $b \in I$ such that $c = ba_{r+1}^{-1} \in I_{r+1}$, $c \notin I$. Now consider

$$Z_0 = \langle x_1, x_2, \ldots, x_r \rangle$$

and

$$Z = \langle x_1 + cx_{r+1}, x_2, \ldots, x_r \rangle .$$

By 14.5 $\varphi_I(Z_0)$ and $\varphi_I(Z)$ are two distinct elements of $S(\tilde{X}, \tilde{Y})$. This shows that in the case $r < r''$ the intersection is not unique.

Let $\dim H_I \geq 3$. A point in H_I is an element of dimension 1, a line an element of dimension 2. Two points are called neighbour points if there exists more than one line joining them. Two lines are called neighbour lines if there exists more than one point contained in both.

COROLLARY 16.3. 1.) The neighbour relation defined on the set of points of H_I is an equivalence relation induced by $\varphi_{P_1(I),I}$ (This means that two points are neighbour points if and only if they are mapped onto the same point by $\varphi_{P_1(I),I}$)

2.) The neighbour relation on the set of lines of H_I is an equivalence relation induced by $\varphi_{P_r(I),I}$.

PROOF : Let $\tilde{X}$ and $\tilde{Y}$ be two distinct points in H_I. Then $\dim V(\tilde{X}, \tilde{Y}) = 2$ by the dimension formula. Their join is different from the whole space because we assume $\dim H_I \geq 3$. Hence, by the first part of 16.2, if their images are distinct under the mapping $\varphi_{P_l(I),I}$ then they have a unique join and are therefore not neighbour points. On the other hand, if their images are equal, the condition on the dimensions is not satisfied. Hence they do not have a unique join and are therefore neighbour points.

The second assertion follows similarly from the second part of 16.2.

Klingenberg [1] calls a Hjelmslev space a geometry with neighbour relation, when the neighbour relation on the set of points and on the set of lines is induced by the residue class mapping.

COROLLARY 16.4. The projective Hjelmslev space H_I is a geometry with neighbour relation if and only if $P_r(I) = P_l(I) = M$.

PROOF : If I satisfies the condition, then from 16.3 follows that φ_{MI} induces the neighbour relation on the set of points and lines.

Suppose that $P_r(I)$ or $P_l(I)$ is properly contained in M, say $a \in M$, $a \notin P_l(I)$. Consider two normally independent vectors $x, u \in T$ such that $\|x\| = \|u\| = |1|$. By 16.2 the points $X = \varphi_I(\langle x\rangle)$ and $Y = \varphi_I(\langle x + au\rangle)$ have a unique join though $\varphi_{MI}(X) = \varphi_{MI}(Y)$. Similarly one concludes from $P_r(I) \neq M$ that φ_{MI} does not induce the neighbour relation on the set of lines.

THEOREM 16.5. Let $\dim H_I \geq 3$. Then H_I is a projective space if and only if I is completely prime.

PROOF : First assume that I is completely prime. Then by 15.2 $P_l(I) = P_r(I) = I$, thus the two conditions of 16.2 hold for all elements of H_I. If the join is the whole space, then it is clearly unique. The same holds for the intersection if this is the null space. Hence join and intersection are unique by 16.2.

Conversely, let I be not completely prime. Then by 15.2 we have $I \neq P_1(I)$, say $a \in P_1(I)$, $a \notin I$. We consider the two 1-dimensional subspaces of T

$$X = \langle x \rangle \quad \text{and} \quad Y = \langle x + w \rangle$$

where x and w are normally independent vectors of T and $\|x\| = |1|$, $\|w\| = |a|$. The join of X and Y is not the whole space because dim T $\geq$ 3. Further, the condition in first part of 16.2 is violated since r=0 and r'=1 (where r and r' have the same meaning as in the proof of 16.2). Hence the join of $\varphi_I(X)$ and $\varphi_I(Y)$ is not unique in H_I.

EXERCISES

1.) Let K be a commutative field and R the 3-dimensional algebra with the basis {1, x, y} satisfying the relations $x^2 = xy = y^2 = 0$. Consider the projective Hjelmslev plane H(T) where T is the free R-module of rank 3 and the points Ru and Rv with u = (0, 0, 1) and v = (x, y, 1). Show that there is no direct summand other than T containing Ru and Rv. (This shows that the dimension formula is not valid in H(T).)

2.) Let T be the free module over a local ring of rank 4 and H(T) the projective Hjelmslev space of T. Show that the dimension formula is equivalent to the following statements:

(1) Two points are joint by a line.

(2) Two lines of a plane intersect in a point.

(3) If two lines intersect in a point then they are contained in a plane.

(4) A line and a plane have a point in common.

(5) If a point and a line are given then there is a plane containing both.

(6) Two planes intersect in a line.

3.) Let a given valuation on K be a V-valuation and I a non-trivial ideal of the valuation ring B. Let $\tilde{X}$, $\tilde{Y}$ be elements of H_I and $\tilde{Z}$ an arbitrary element of $V(\tilde{X}, \tilde{Y})$. Then there exist subspaces X and Y of T and a non-zero ideal I′ contained in I such that all elements of $V(\varphi_{I'}(X), \varphi_{I'}(Y))$ are mapped onto $\tilde{Z}$ by $\varphi_{I'I}$. An analogous result holds for the intersection set.

4.) Let $\tilde{X}, \tilde{Y} \in H_I$ where X and Y are defined as in exercise 4 of chapter 3. Determine $V(\tilde{X}, \tilde{Y})$ and $S(\tilde{X}, \tilde{Y})$.

5.) Show for an ideal I of a valuation ring B

$$aI < I \iff a \in P_l(I)$$

$$Ia < I \iff a \in P_r(I),$$

where < means the proper inclusion.

APPENDIX: LOCALLY INVARIANT VALUATIONS

by J.Graeter

In this appendix we ask, under which circumstances the general approximation theorem (in the sense of Ribenboim [2]) holds for valuations of skew fields. We introduce the special class of locally invariant valuations. Essentially, only these valuations satisfy the general approximation theorem.

Let B be a valuation ring of K and M the maximal ideal of B. For all $x \in M$ denote by $P(x)$ the minimal completely prime ideal of B containing x. B is **locally invariant** if

$$xP(x) = P(x)x$$

holds for all $x \in M$. A valuation of K is locally invariant if the corresponding valuation ring is so.

(A) Invariant valuation rings are locally invariant: Let B be invariant with the maximal ideal M and $x \in M$, $x \neq 0$. Since $xBx^{-1} = B$, $xP(x)x^{-1}$ is a completely prime ideal of B containing x. By the minimality of $P(x)$ $xP(x)x^{-1}$ includes $P(x)$. $P(x) \subseteq x^{-1}P(x)x$ follows similarly. This means $P(x) = xP(x)x^{-1}$.

(B) A maximal locally invariant valuation ring B is invariant (B is maximal if B has no proper overring in K): By assumption, the maximal ideal M of B and {0} are the only completely prime ideals of B. Thus, $P(x) = M$ holds for all $x \in M$, $x \neq 0$. Since B is locally invariant, each $x \in K$ with $x \notin B$ satisfies $x^{-1}M = Mx^{-1}$ and each $x \in B$ satisies $xM = Mx$. Clearly M is invariant and thereby B.

(C) Let P be a completely prime ideal of the locally invariant valuation ring B and B′ the corresponding valuation ring (see 1.7). The factor ring B/P is a locally invariant valuation ring of B′/P: Obviously, B/P is a valuation ring of B′/P. B/P is locally invariant because for each nonunit $x \in B\setminus P$, $P(x)/P$ is the minimal completely prime ideal of B/P containing $x + P$.

(D) Each overring of a locally invariant valuation ring is locally invariant: This follows directly by definition.

(E) Locally invariant valuations are V-valuations: Let I be a non-zero right ideal of the locally invariant valuation ring B and $x \in I$, $x \neq 0$. Since $xP(x) = P(x)x$ holds, $xP(x)$ is an ideal of B. I contains $xP(x)$. By 6.3, the valuation belonging to B is a V-valuation.

(F) Examples of locally invariant valuations, which are not invariant, can be constructed by valuations of formal power series fields. Let G be a totally ordered group, K a field and K(G) the formal power series field. In section 4, it was shown that each valuation $|\ |: K \to W$ can be extended to a valuation $|\ |': K(G) \to W'$. It is easily checked, that $|\ |'$ is locally invariant if $|\ |$ is locally invariant. In particular, $|\ |'$ is locally invariant if K is commutative. Hence the first example of section 4 is locally invariant, but not invariant.

Let R be a ring and I a left or right ideal of R. The subset Rad(I), consisting of all $x \in R$ with $x^k \in I$ for some positive integer k, is called the **radical** of I. Generally, the radical of a left (resp. right) ideal of a noncommutative ring is not a left (resp. right) ideal.

PROPOSITION 1. If B is a valuation ring of K, having the maximal ideal M and $|\ |$ a corresponding valuation, then the following statements are equivalent:

(1) $|\ |$ is locally invariant.

(2) For each right ideal I of B, Rad(I) is a completely prime ideal of B.

(3) For each $x \in M$, $\{ y \in B \mid |y| \leq |x^k| \text{ for all positive integers } k \}$ is a completely prime ideal of B.

PROOF. (1) $\Rightarrow$ (2): Let P_1 be the maximal completely prime ideal of B with $P_1 \subseteq I$ and P_2 the minimal completely prime ideal of B containing I. The case $I = P_1$ is trivial. Thus, let $I \neq P_1$. B_i denotes the valuation ring of K belonging to P_i. By (C), B/P_1 is a locally invariant valuation ring of B_1/P_1. B_2/P_1 is a maximal valuation ring of B_1/P_1 containing B/P_1. Therefore, B_2/P_1 is locally invariant (see (D)) and evenly invariant (see (B)). Since $I \neq P_1$, there exists a non-zero right ideal A of B_2/P_1 with $A \subseteq I/P_1$. B_2/P_1 is maximal and invariant, thus we have

$$P_2/P_1 = \mathrm{Rad}(A) \subseteq \mathrm{Rad}(I/P_1) = \mathrm{Rad}(I)/P_1 \subseteq P_2/P_1.$$

Clearly, $P_2/P_1 = \mathrm{Rad}(I)/P_1$ and $P_2 = \mathrm{Rad}(I)$ is valid.

(2) ⇒ (3): Let $x \in M$, $A := \{ y \in B \mid |y| \leq |x^k|$ for all positive integers $k \} = xB \cap x^2B \cap \ldots \cap x^kB \cap \ldots$ and P the maximal completely prime ideal of B with $P \subseteq A$. Suppose $P \neq A$. Then there exists $y \in A$ with $y \notin P$. By (2), $Rad(yB)$ is a completely prime ideal of B minimal with $yB \subseteq Rad(yB)$. As an intersection of right ideals, A is a right ideal. Therefore, we have $A \subseteq Rad(yB)$ or $Rad(yB) \subseteq A$. By the maximality of P and $P \neq Rad(yB)$, $Rad(yB) \subseteq A$ is false. Thus, $A \subseteq Rad(yB)$, $A \neq Rad(yB)$ and $Rad(yB) \subseteq x^kB$ cannot hold for each positive integer k. Therefore, $x \in Rad(yB)$, hence $x^n \in yB$ and $A \subseteq x^nB \subseteq yB$ for some positive integer n. This is a contradiction since $A \neq x^nB$ and $y \in A$.

(3) ⇒ (1): Let $x \in M$, $x \neq 0$. For each $y \in P(x)$ there exists positive integer k with $|y^k| < |x^2|$, otherwise $\{ z \in B \mid |z| \leq |y^k|$ for all positive integers $k \}$ would be a smaller completely prime ideal than $P(x)$ containing x. $|(x^{-1}yx)^k| = |x^{-1}y^kx| < |x^{-1}y^k| < |x|$. Thus, $x^{-1}yx \in P(x)$ and $x^{-1}P(x)x \subseteq P(x)$. This shows that $x^{-1}P(x)x$ is a completely prime ideal of B containing x. The minimality of $P(x)$ implies $x^{-1}P(x)x = P(x)$.

By symmetry we have:

PROPOSITION 2. A valuation ring B of K is locally invariant if and only if $Rad(I)$ is a completely prime ideal for each left ideal I of B.

THEOREM 3. A valuation ring B of K is locally invariant if the following is valid:
For each $x \in K$ there exists a non-zero integer k with

$$Bx^k \subseteq x^kB.$$

For the proof we need:

PROPOSITION 4. Let B be a valuation ring of K and I an ideal of B. Then,

$$A := \bigcap_{k=1}^{\infty} I^k$$

is a completely prime ideal of B.

PROOF. Obviously, A is an ideal. Let $x,y \in B$ and $xy \in A$. Suppose $x \notin I^k$ for some positive integer k. Then $I^k \subseteq xB$ and $x^{-1}I^k \subseteq B$. For all positive integers n we have $xy \in A \subseteq I^kI^n$ and therefore $y \in x^{-1}I^kI^n \subseteq BI^n \subseteq I^n$. This means $y \in A$.

Now we can prove Theorem 3.

PROOF. Let I be an arbitrary right ideal of B, P the minimal completely prime ideal of B containing I and $y \in P$. $x := y^2$. Showing $x^n \in I$ for a positive integer n proves $Rad(I) = P$. By assumption, there exists a positive integer k with $Bx^k \subseteq x^kB$ or $x^kB \subseteq Bx^k$. Let $Bx^k \subseteq x^kB$. The case $x^kB \subseteq Bx^k$ is similar. $x^kB \subseteq Bx^kB \subseteq x^kB$ implies that $A = x^kB$ is an ideal. By Proposition 4, $D := A \cap A^2 \cap \ldots \cap A^n \cap \ldots$ is a completely prime ideal of B. $D \subseteq x^kB \subseteq yB \subseteq P$ and $x^kB \neq yB$ gives $D \neq P$. Since P is the minimal completely prime ideal of B containing I, I includes D properly. Hence, $x^n \in I$ for some positive integer n.

COROLLARY 5. A valuation ring B of K is locally invariant if B contains an invariant valuation ring B′ of K.

PROOF. For all $x \in K$, $x \neq 0$; $B' = xB'x^{-1} \subseteq xBx^{-1}$ is valid. Since xBx^{-1} and B are overrings of B′, 1.3. implies $xBx^{-1} \subseteq B$ or $B \subseteq xBx^{-1}$.

COROLLARY 6. Let L be a field algebraic over its centre K. Each valuation $|\ |$ of L is locally invariant.

PROOF. Let $x \in L$, $x \neq 0$. Similar to the commutative case, there exists a positive integer n and $k \in K$ satisfying $|x^n| = |k|$. If B is the valuation ring of K belonging to $|\ |$, for each $b \in B$ $|bx^n| = |bk| = |kb| \leq |k| = |x^n|$ holds. Therefore, $Bx^n \subseteq x^nB$ is valid.

COROLLARY 7. A valuation ring B of K is locally invariant if B has only a finite number of conjugate valuation rings.

PROOF. Straightforward.

Let $|\ |_1, |\ |_2$ be valuations of K ($|\ |_i : K \to W_i$) with the valuation rings B_1, B_2. The minimal valuation ring containing B_1 and B_2 is denoted by $B_{1\cdot 2}$ and the corresponding valuation by $|\ |_{1\cdot 2}$. Clearly, $|\ |_1$ and $|\ |_2$ are comaximal if and only if $B_{1\cdot 2} = K$ holds. If $M_{1\cdot 2}$ is the maximal ideal of $B_{1\cdot 2}$, $B_1/M_{1\cdot 2}$ and $B_2/M_{1\cdot 2}$ are comaximal valuation rings of $B_{1\cdot 2}/M_{1\cdot 2}$. Now we give some definitions which will be used in this section.

(1) A pair $(\varepsilon_1, \varepsilon_2) \in W_1^\times \times W_2^\times$ is called **compatible** if there exist $a, b \in K$ satisfying

$$|a|_1 = \varepsilon_1,\ |b|_2 = \varepsilon_2 \text{ and } |a|_{1\cdot 2} = |b|_{1\cdot 2}.$$

(2) $(\varepsilon_1, \varepsilon_2, a_1, a_2) \in W_1^\times \times W_2^\times \times K^2$ is called compatible if $(\varepsilon_1, \varepsilon_2)$ is compatible and there exist $a, b \in K$ satisfying

$$|a|_1 = \varepsilon_1,\ |b|_2 = \varepsilon_2 \text{ and } |a|_{1\cdot 2} = |b|_{1\cdot 2} \geq |a_1 - a_2|_{1\cdot 2}.$$

Obviously, if $|\ |_1$ and $|\ |_2$ are comaximal, each $(\varepsilon_1, \varepsilon_2, a_1, a_2)$ is compatible.

PROPOSITION 8. Let $|\ |_1, |\ |_2$ be valuations of K with the corresponding value sets W_1, W_2, the valuation rings B_1, B_2 and the maximal ideals M_1, M_2. If $|\ |_1$ is locally invariant, for each compatible $(\varepsilon_1, \varepsilon_2, a_1, a_2) \in W_1^\times \times W_2^\times \times K^2$ there exists an $x \in K$ such that

$$|x - a_1|_1 \leq \varepsilon_1,\ |x - a_2|_2 \leq \varepsilon_2$$

holds.

PROOF. $R := B_1 \cap B_2$. First, let $|\ |_1, |\ |_2$ be comaximal. Without restrictions we assume, that $B_i \neq K$ and $\varepsilon_i < |1|_i$ holds.

$$A_i := \{\ x \in R \mid |x|_i \leq \varepsilon_i\ \}$$

$$Q_i := \{\ x \in K \mid |x^k|_i \leq \varepsilon_i \text{ for some positive integer } k\ \}$$

Since $|\ |_1$ is locally invariant, Q_1 is a completely prime ideal of B_1. We distinguish three cases:

(1) $A_1 + A_2 \subseteq R \cap M_2$.

B denotes the valuation ring of K belonging to Q_1.
$S := \{ r \in R \mid r \notin Q_1 \}$. 7.4. shows $B = RS^{-1}$. Since $R \cap Q_1 \subseteq R \cap M_1 \cap M_2$, $B_1, B_2 \subseteq B$ holds. This contradicts the comaximality of B_1 and B_2. Thus, case (1) cannot occur.

(2) $A_1 + A_2 \subseteq R \cap M_1$.
For $x \in A_2$, $x \neq 0$ denote by $P(x)$ the minimal completely prime ideal of B_1 containing x and by $|\ |_1'$ the corresponding valuation of K. 7.4. and the comaximality of B_1 and B_2 show, that $M_2 \cap R$ cannot contain $P(x) \cap R$. Choose $y \in R$ with $|y|_2 = |1|_2$ and $y \in P(x)$. Since $|\ |_1'$ is locally invariant, there exists a positive integer k, such that $|y^k|_1' < |x|_1'$. Therefore, $x^{-1}y^k \in P(x) = x^{-1}P(x)x$ and $y^k x^{-1} \in P(x)$ holds. Then we have $|y^k x^{-1}|_1 < |1|_1$ and $|y^k x^{-1}|_2 > |1|_2$. Thus, $|(1+y^k x^{-1})^{-1}|_1 = |1|_1$ and $|1+y^k x^{-1}|_2 = |y^k x^{-1}|_2$ is valid. $y^k x^{-1}(1+y^k x^{-1}) = (1+y^k x^{-1})y^k x^{-1}$ implies $|(1+y^k x^{-1})(y^k x^{-1})^{-1}|_2 = |1|_2$ and $|(1+y^k x^{-1})^{-1}|_2 = |(y^k x^{-1})^{-1}|_2 = |x(y^{-1})^k|_2 = |x|_2 \leq \varepsilon_2$. Finally, $(1+y^k x^{-1})^{-1} \in A_2$ and $(1+y^k x^{-1})^{-1} \notin M_1$ contradict our assumption. Therefore, case (2) cannot occur.

(3) There exist $x_1, x_2 \in R$ with $x_i \in A_1 + A_2$, $x_i \notin M_i$.
If $x_1 \notin M_2$ or $x_2 \notin M_1$ is valid, then x_1 or x_2 is a unit and therefore, $A_1 + A_2 = R$. Otherwise, $x_1 + x_2$ is a unit. In any case, we have $A_1 + A_2 = R$. Now we show the existence of an $x \in K$ satisfying $|x-a_1|_1 \leq \varepsilon_1$, $|x-a_2|_2 \leq \varepsilon_2$. If $a_1, a_2 \in R$, $a_1 - a_2 = r_1 + r_2$ holds for some suitable $r_1, r_2 \in R$, $r_i \in A_i$. Put $x = a_1 - r_1 = r_2 + a_2$. If $a_1 \notin R$ or $a_2 \notin R$, then, by 7.9., there exist $b_1, b_2, b \in R$, $b \neq 0$ satisfying $a_1 = b^{-1}b_1$, $a_2 = b^{-1}b_2$. Choose y_1, $y_2 \in K$ with $|b^{-1}y_i|_i = \varepsilon_i$. There exists an $x \in R$ such that $|x-b_1|_1 \leq |y_1|_1$, $|x-b_2|_2 \leq |y_2|_2$. Therefore, $|b^{-1}x-a_1|_1 \leq \varepsilon_1$, $|b^{-1}x-a_2|_2 \leq \varepsilon_2$ holds.
Now, we consider the case when $|\ |_1$ and $|\ |_2$ are not necessarily comaximal. Let $|\ |_{1\cdot 2}$, $B_{1\cdot 2}$ and $M_{1\cdot 2}$ be defined as above. $|\ |_i'$ denotes the valuation of $B_{1\cdot 2}/M_{1\cdot 2}$ corresponding to the valuation ring $B_i/M_{1\cdot 2}$. $|\ |_1'$ is locally invariant. Choose $y_1, y_2 \in K$ with $|y_i|_i = \varepsilon_i$. If $|y_1|_{1\cdot 2} = |y_2|_{1\cdot 2} > |a_1-a_2|_{1\cdot 2}$ holds, then put $x = a_2$ and $|x-a_1|_1 \leq \varepsilon_1$; $|x-a_2|_2 \leq \varepsilon_2$ is valid. Therefore, assume $|y_1|_{1\cdot 2} = |y_2|_{1\cdot 2} = |a_1-a_2|_{1\cdot 2}$. $z_i := (a_1-a_2)^{-1}y_i$. $|z_i|_{1\cdot 2} = |1|_{1\cdot 2}$. By the comaximality of $|\ |_1'$ and $|\ |_2'$ there exists $y \in B_{1\cdot 2}$, such that

$$|\bar{1}-\bar{y}|_1' \leq |\bar{z}_1|_1', \quad |\bar{y}|_2' \leq |\bar{z}_2|_2'$$

holds. Consequently,

$$|1-y|_1 \leq |z_1|_1, \quad |y|_2 \leq |z_2|_2$$

is valid. Finally, we have

$$|(a_1-a_2)y-(a_1-a_2)|_1 \leq |y_1|_1, \quad |(a_1-a_2)y|_2 \leq |y_2|_2.$$

Put $x = (a_1-a_2)y+a_2$.

Using 7.10 we get the main result of this section:

THEOREM 9. (General approximation theorem)

Let $| \ |_1, \ldots, | \ |_n$ be valuations of K. If $| \ |_2, \ldots, | \ |_n$ are locally invariant, for each $(\varepsilon_1, \ldots, \varepsilon_n, a_1, \ldots, a_n) \in W_1^\times \times \ldots \times W_n^\times \times K^n$ there exists an $x \in K$ satisfying

$$|x-a_1|_1 \leq \varepsilon_1, \ \ldots \ , |x-a_n|_n \leq \varepsilon_n,$$

if each $(\varepsilon_i, \varepsilon_k, a_i, a_k)$ is compatible ($i,k = 1,\ldots,n$).

Finally, we will show that the general approximation theorem fails, if two of the valuations are arbitrary.

PROPOSITION 10. Let $| \ |_1$ be a non-locally-invariant valuation of K. Then there exists a valuation $| \ |_2$ of K and a compatible $(\varepsilon_1, \varepsilon_2, a_1, a_2) \in W_1^\times \times W_2^\times \times K^2$, such that no $x \in K$ satisfies

$$|x-a_1| \leq \varepsilon_1, \ |x-a_2|_2 \leq \varepsilon_2.$$

PROOF. For each $x \in M_1$, $x \neq 0$ denote the minimal completely prime ideal of B_1 containing x by $P(x)$. Suppose $xP(x)x^{-1} \neq P(x)$ for such an x. Let $| \ |_2$ be the valuation of K belonging to xB_1x^{-1}. Since $x \in xP(x)x^{-1}$, by the minimality of $P(x)$, $xP(x)x^{-1} \subsetneq P(x)$ cannot hold. $x^{-1}P(x)x \subsetneq P(x)$ and therefore $P(x) \subsetneq xP(x)x^{-1}$ fails similarly. Thus, $M_{1\cdot 2} \neq P(x)$ is valid and $|x|_{1\cdot 2} = |1|_{1\cdot 2}$. Clearly $(|x^2|_1, |x|_2, x, 1)$ is compatible. Suppose: There exists a $y \in K$ with

$$|y-x|_1 \leq |x^2|_1, \ |y-1|_2 \leq |x|_2.$$

$x \in M_1$ implies $x \in xM_1x^{-1} = M_2$. Thus, we have $|y-x|_1 \leq |x^2|_1 < |x|_1$ and $|y-1|_2 \leq |x|_2 < |1|_2$. Therefore, $|y|_1 = |x|_1$ and $|y|_2 = |1|_2$ holds. Hence, $x^{-1}yx$ is a unit in B_1. Finally, $x^{-1} = (x^{-1}y^{-1}x)(x^{-1}y) \in B_1$ contradicts $x \in M_1$.

EXERCISES

1.) Let $| \ |$ be a valuation of K and let B be the correspon-ding valuation ring. Let $U_0 \subseteq U_1 \subseteq \ldots \subseteq U_n$ be all the dis-tinct φ-convex subgroups of the value group of $| \ |$. Show that the following statements are equivalent:

(1) $| \ |$ is locally invariant.

(2) U_i is normal in U_{i+1} ($i < n$).

2.) Let $| \ |_1, | \ |_2$ be distinct valuations of K and let W_1, W_2 be the corresponding value sets. Prove: if each compatible $(\varepsilon_1, \varepsilon_2, a_1, a_2) \in W_1^{\times} \times W_2^{\times} \times K^2$ has an $x \in K$ satisfying

$$|x-a_1|_1 \leq \varepsilon_1, \quad |x-a_2|_2 \leq \varepsilon_2$$

then

(a) each compatible $(\varepsilon_1, \varepsilon_2) \in W_1^{\times} \times W_2^{\times}$ possesses an $x \in K$, such that $|x|_1 = \varepsilon_1$, $|x|_2 = \varepsilon_2$ holds.

(b) each compatible $(\varepsilon_1, \varepsilon_2, a_1, a_2) \in W_1^{\times} \times W_2^{\times} \times K^2$ possesses an $x \in K$, such that $|x-a_1|_1 = \varepsilon_1$, $|x-a_2|_2 = \varepsilon_2$ holds.

3.) Let $| \ |_1, | \ |_2$ be valuations of K and let W_1, W_2 be the corresponding value sets. Prove: for each compatible $(\varepsilon_1, \varepsilon_2, a_1, a_2) \in W_1^{\times} \times W_2^{\times} \times K^2$ there exists an $x \in K$ satisfy-ing

$$|x-a_1|_1 \leq \varepsilon_1, \quad |x-a_2|_2 \leq \varepsilon_2,$$

if each compatible $(\varepsilon_1, \varepsilon_2) \in W_1^{\times} \times W_2^{\times}$ possesses an $x \in K$, such that

$$|x|_1 = \varepsilon_1, \quad |x|_2 = \varepsilon_2$$

holds.

SOLUTIONS

Exercises of Chapter 1

1.) Let I be a fractional left or right ideal. Then

$$(aI)^* = \{b \in K \mid 1 \notin baI\} = \{b \in K \mid ba \in I^*\} = I^*a^{-1}.$$

By 1.5 $B^* = M$, hence $(aB)^* = Ma^{-1}$.

2.) a) Let $I = I(H) = \{a \in K \mid |a| \in H\}$. Since $H \neq \emptyset$ we have $0 \in H$, hence $0 \in I$. Let $a \in I$, $b \in B$. From $|ab| \leq |a| \in H$ it follows $|ab| \in H$, hence $ab \in I$. Conversely, let I be a fractional right ideal. Define $H = H(I) = \{|a| \mid a \in I\}$. Then H is a lower class and $I = IH(I)$ and $H = HI(H)$.

b) The composition of the star-mapping and the mapping $H \longmapsto I(H)$ is bijective. Then we have

$$a \in I(H)^* \Leftrightarrow a^{-1} \notin I(H) \Leftrightarrow |a^{-1}| \notin H \Leftrightarrow |1| \notin v(a)H$$

$$\Leftrightarrow v(a)h \leq |1| \text{ for all } h \in H.$$

3.) (1) $\Leftrightarrow$ (2). Let $a \in M$. The invariance of B implies $bab^{-1} \in B$. Because E is invariant by 2.5 and $a \notin E$, bab^{-1} is not a unit. Conversely, let M be invariant and $a \in B$. Suppose $bab^{-1} \notin B$, then $ba^{-1}b^{-1} \in M$, hence $a^{-1} \in bMb^{-1} \subseteq M$. From this it follows $a \notin B$, a contradiction.

(1) $\Leftrightarrow$ (3). Let B be invariant and I a right ideal of B and $a \in I$, $a \neq 0$. Then $Ba = a(a^{-1}Ba) = aB \subseteq I$, hence I is a two-sided ideal. Similarly, I is shown to be two-sided when I is left ideal. Conversely, let all right and left ideals of B be two-sided. For $a \in B$ we have $aB \subseteq BaB \subseteq Ba$ and $Ba \subseteq BaB \subseteq aB$, hence $aB = Ba$. If $a \notin B$ we have $a^{-1} \in B$, hence $a^{-1}B = Ba^{-1}$. In any case $B = a^{-1}Ba$ holds.

4.) (1) $\Rightarrow$ (2). $|a| = |b|$ implies $|a^{-1}b| = |1|$, hence $a^{-1}b \in E$. By 2.5 E is invariant, hence $ba^{-1} = a(a^{-1}b)a^{-1} \in E$, hence $|a^{-1}| = |b^{-1}|$.

(2) $\Rightarrow$ (3). Let $a, b, c \neq 0$. Then $|a| = |b|$ implies $|a^{-1}| = |b^{-1}|$, hence $|(ac)^{-1}| = |c^{-1}a^{-1}| = |c^{-1}b^{-1}| = |(bc)^{-1}|$, hence $|ac| = |bc|$.

(3) $\Rightarrow$ (4). $|ab| = |1|$ implies $|a| = |abb^{-1}| = |b^{-1}|$ by (3), hence $|ba| = v(b)|a| = v(b)|b^{-1}| = |1|$.

(4) $\Rightarrow$ (1). Let $a \in E$, $b \in K^\times$. From $|(ab)b^{-1}| = |a| = |1|$ it follows $|b^{-1}ab| = |1|$. Hence E and therefore B are invariant.

5.) (1) ⇒ (2). Let B_1 be an invariant valuation ring contained in B. Then the conjugate valuation rings bBb^{-1}, $b \in K^\times$, are fractional ideals with respect to B_1. By 1.2 they form a totally ordered set.

(2) ⇒ (3). Let B_0 be the intersection of all valuation rings bBb^{-1}, $b \in K^\times$. Obviously, B_0 is invariant. Further, $B_0 = \{ a \in K \mid v(a) \leq v(1) \}$ because $v(a) \leq v(1)$ is equivalent to $|ab| \leq |b|$ or $|b^{-1}ab| \leq |1|$ or $a \in bBb^{-1}$ for all $b \in K^\times$, hence $a \in B_0$.

If the valuation rings bBb^{-1}, $b \in K^\times$, form a totally ordered set, then B_0 is a valuation ring. Consequently, for a, $b \in K^\times$ we have $a^{-1}b \in B_0$ or $b^{-1}a \in B_0$, hence $v(a^{-1}b) \leq v(1)$ or $v(b^{-1}a) \leq v(1)$, hence $v(b) \leq v(a)$ or $v(a) \leq v(b)$. This shows that G is totally ordered.

(3) ⇒ (1). If G is totally ordered and $a \in K^\times$, then $v(a) \leq v(1)$ or $v(1) \leq v(a)$, hence $a \in B_0$ or $a^{-1} \in B_0$. Therefore, B_0 is an invariant valuation ring contained in B.

Exercises of Chapter 2

1.) Let I be a non-zero right ideal of B and v a non-zero element of I. Let aB be a bounded neighbourhood of 0. Then there exists $b \in K^\times$ such that $aBb \subseteq avB$ which implies $Bb \subseteq vB \subseteq I$. Hence Bb is a non-zero left ideal contained in I. By 5.3 B generates a V-topology.

2.) First assume that the condition holds. Let the non-zero right ideal vB be given. Then there exists $\varepsilon' \in W^\times$ such that $|a| > \varepsilon'$ implies $|a^{-1}| \leq |v|$ for all $a \in K^\times$. Choose $u \in K^\times$ with $|u^{-1}| = \varepsilon'$. Let $a \in Mu$, $a \neq 0$. Since $Mu = (Cu^{-1}B)^{-1} \cup \{0\}$, we have $a^{-1} \notin u^{-1}B$, hence $|a^{-1}| > \varepsilon' = |u^{-1}|$. The condition implies $|a| \leq |v|$ or $a \in vB$. This shows $Mu \subseteq vB$. By 6.3 the valuation is a V-valuation.

Conversely, let $\varepsilon = |v| > 0$. If the valuation is a V-valuation, by 6.3 there exists $u \in K^\times$ such that $Bu \subseteq vB$. Put $\varepsilon' = |u^{-1}|$. When $|a| > \varepsilon'$ we have $a \in Cu^{-1}B$, hence $a^{-1} \in Mu \subseteq Bu \subseteq vB$ and therfore $|a^{-1}| \leq \varepsilon = |v|$.

3.) First choose $\varepsilon_i' \in W_i^\times$ with $\varepsilon_i' < \varepsilon_i$ and $a_i \in K$ such that $|a_i| = \varepsilon_i$ for i = 1, ... , n. By 8.2 there exists an element $a \in K$ such that $|a - a_i|_i \leq \varepsilon_i' < \varepsilon_i = |a_i|_i$. By the principle of domination we have $|a|_i = |a_i + (a-a_i)| = |a_i|_i = \varepsilon_i$ for i = 1, ... , n.

4.) From $a \in P$ it follows $(aB_1)B_1 \subseteq aB_1 \subseteq M_2$ and $(B_2a)B_1 = B_2(aB_1) \subseteq B_2M_2 \subseteq M_2$, hence $PB_1 \subseteq P$ and $B_2P \subseteq P$. Further, P is a proper left ideal of B_2 because $P \subseteq PB_1 \subseteq M_2$. In order to show $P \subseteq M_1$ suppose $a \in P$, $a \notin M_1$. The latter implies $a^{-1} \in B_1$, hence $1 = aa^{-1} \in M_2$ a contradiction.

If the topologies generated by B_1 and B_2 are equal, M_2 is a neighbourhood of 0 with respect to the topology generated by B_1, hence there exists $a \in K^{\times}$ such that $aB_1 \subseteq M_2$, hence $P \neq 0$.

5.) Let the valuations be independent and suppose $B_1aB_2 \neq K$ for some $a \in K^{\times}$. Let $b \in K^{\times}$, $b \notin B_1aB_2$. By the independence of the valuations there exists $c \in K$ such that

$$|c|_1 = |a^{-1}|_1 \quad \text{and} \quad |c|_2 = |1|_2.$$

Hence ac is a unit in B_1, c a unit in B_2. Therefore $B_1 = B_1ac \subseteq B_1aB_2$. Further, there exists $d \in K$ such that

$$|d|_1 = |1|_1 \quad \text{and} \quad |d|_2 = |b|_2$$

which implies

$$b \in bB_2 = dB_2 \subseteq B_1B_2 \subseteq B_1aB_2B_2 \subseteq B_1aB_2$$

contrary to the choice of b.

Conversely, let $B_1aB_2 = K$ for all $a \in K^{\times}$. We may take over the second part of the proof of 8.1 ((2) $\Rightarrow$ (3)). We must only show that $B_1b_2B_2 = K$ implies $I_2S^{-1} = B_1$. Indeed, we have $uv = 1$ with $u \in B_1$, $v \in b_2B_2$. Put $s = v$ if $|u|_1 = |1|_1$ and $s = v(1+v)^{-1} = (1+u)^{-1}$ if $|u|_1 < |1|_1$. In each case we have $|s|_1 = |1|_1$ and $|s|_2 \leq |b_2|_2$, hence $s \in I_2$ and $s \in S$. This implies $I_2S^{-1} = B_1$ because $1 = ss^{-1} \in I_2S^{-1}$.

6.) By exercise 5 $B_1aB_2 = K$ holds for all $a \in K^{\times}$. From this it follows $B_1b(cB_2c^{-1}) = K$ for all $b, c \in K^{\times}$. Now apply exercise 5 again.

7.) The intersection of all non-zero completely prime ideals is also a completely prime ideal. Because of the correspondence between overrings and completely prime ideals (conf. 1.7) the intersection must be the null ideal. Hence the set of all non-zero completely prime ideals is a fundamental system of neighbourhoods of 0. Hence the topology is induced by two-sided ideals and is therefore a V-topology.

8.) By exercise 4 $P = \{ a \in K \mid aB_1 \subseteq M_2 \}$ is a right ideal of B_1. Suppose $P \neq 0$. Since the non-zero completely prime ideals of an infinite valuation ring form a filter base, there exists a non-zero completely prime ideal $P_1 \subseteq P$ in B_1. By 1.6 P_1 satisfies the condition (P). Further, by exercise 4 P_1 is contained in M_2. Hence, by 1.6 P_1 is also a completely prime ideal of B_2. This contradicts the assumption that B_1 and B_2 are comaximal since P_1 corresponds to a common overring of B_1 and B_2 by 1.7. Therefore we obtain $P = 0$. Hence the topology induced by $|\ |_1$ is not finer than the topology induced by $|\ |_2$. Since $|\ |_1$ is a V-valuation by exercise 7, the valuations are independent by 8.6.

9.)Let $a \in K$ be analytically nilpotent and P a non-zero completely prime ideal. Then $a^n \in P$ which implies $a \in P$ because P is completely prime. Therfore, N is contained in the intersection of all non-zero completely prime ideals. If the valuation is infinite the intersection is null, hence $N = 0$.

Exercises of Chapter 3

1.) Let $y \in U$. Since x_0 is a best approximation of x in U we have $\|x - x_0\| \leq \|x - u\|$ for all $u \in U$, hence

$$\|x - x_0\| \leq \|(x - x_0) + y\| \leq \mathrm{Max}(\|x - x_0\|, \|y\|).$$

If $\|y\| = \|x - x_0\|$ this implies

$$\|(x - x_0) + y\| = \mathrm{Max}(\|x - x_0\|, \|y\|).$$

If $\|y\| \neq \|x - x_0\|$ the same holds by the principle of domination. Therefore, $x - x_0$ and y are normally independent.

2.) First, let T be a valued vector space. Let $\bar{Y}$ be a subspace of $\varphi(U)$ which is complementary to $\varphi(X)$ in $\varphi(U)$. By 9.3 there exists a subspace Y of U with $\varphi(Y) = \bar{Y}$. Obviously, $X + Y \subseteq U$. Further, since the sum is direct, we have $\dim(X + Y) = \dim X + \dim Y$. Since T is a valued vector space, φ preserves the dimension of finite dimensional subspaces. Therefore, we have

$$\dim_K(X + Y) = \dim_{\bar{K}}\varphi(X) + \dim_{\bar{K}}\varphi(Y) = \dim_{\bar{K}}\varphi(U) = \dim_K U$$

which implies $X + Y = Z$.

Conversely, let $y \in T$ and X be a finite dimensional subspace of T. Consider $U = X + \langle y \rangle$. Then there exists a one-dimensional subspace $Y = \langle y_0 \rangle$ of U such that y_0 and X are normall

independent and $U = X + Y$. Let $y = x_0 + ay_0$ with $x \in X$ and $a \in K$. Then for all $x \in X$ we have

$$\|y - x\| = \|y - x_0 + x_0 - x\| = \mathrm{Max}(\|y - x_0\|, \|x_0 - x\|),$$

hence $\|y - x_0\| \leq \|y - x\|$. Therefore, T is a valued vector space.

3.) Clearly, T is a normed vector space. Let U be a subspace of finite dimension and $\{x_1, \ldots, x_k\}$ a basis of U. By elementary operations we obtain a basis $\{y_1, \ldots, y_k\}$, $y_i = (a_{i1}, a_{i2}, \ldots)$, such that $a_{ii} = 1$ and $a_{ik} = 0$ for $i < k$. This is a normal basis of U, hence, φ preserves the dimension of finite dimensional subspaces. Therefore, T is a valued vector space.

Suppose there exists a normal basis of T. Since the residue class space has countable dimension the basis of T is denumerable, say $\{x_1, x_2, \ldots\}$. By the Steinitz exchange process we may assume that the vectors $e_i = (0, \ldots, 1, 0, \ldots)$ occur in this basis. The images of e_i under the residue class homomorphism form a basis of the residue class space. Hence there cannot be any vector other than e_i, $i = 1, 2, \ldots$ in the normal basis considered above. This means that in the representation of the elements $x = (a_1, a_2, \ldots)$ only a finite number of coordinates are non-zero. Consider a sequence $\varepsilon_i \in W^x$ with $\lim \varepsilon_i = 0$ and choose $a_i \in K$ such that $|a_i| = \varepsilon_i$. Then $x = (a_1, a_2, \ldots)$ is an element of T which is not a linear combination of the vectors e_i.

4.) We have $X \cap Y = 0$ and $X + Y = T$. In the residue class space $\varphi(Y)$ is contained in $\varphi(X)$ and has the basis $\{v_0, v_1\}$ with

$$v_0 = (1, 0, 0, 0, 0) \quad \text{and} \quad v_1 = (0, 1, 0, 0, 0).$$

Put $v_2 = (0, 0, 1, 0, 0)$. Then $\{v_0, v_1, v_2\}$ is a basis of $\varphi(X)$. Let Z be the subspace of T spanned by the vectors $(0, 0, 0, 1, 0)$ and $(0, 0, 0, 0, 1)$. Then

$$T = X + Z$$

is a normally independent decomposition of T. Choose inverse images of v_0 and v_1

$$x_0' = (1, t, 0, 0, 0) \quad \text{and} \quad x_1' = (0, 1+t, t, 0, 0) \quad \text{in } X$$

$y_0' = (1, t, 0, t^2, 0)$ and $y_1' = (1, 1+t, t, t^2, t^3)$ in Y

which satisfy the conditions

$$y_0' - x_0' = z_0 = (0, 0, 0, t^2, 0) \in Z$$

$$y_1' - x_1' = z_1 = (0, 0, 0, t^2, t^3) \in Z.$$

Following the method of 11.2 we must determine a set of normally independent generators of the B-module $W = \langle z_0, z_1 \rangle$. In general this is achieved by 11.1. In this example, apparently

$$w_0 = z_1 - z_0 = (0, 0, 0, 0, t^3) \quad \text{and}$$

$$w_1 = z_0 = (0, 0, 0, t^2, 0)$$

form such a system of generators of W. From this we obtain

$$x_0 = x_1' - x_0' = (-1, 1, t, 0, 0) \quad \text{and}$$

$$x_1 = x_0' = (1, t, 0, 0, 0).$$

Putting $x_2 = (0, 0, 1, 0, 0)$ as an inverse image of v_2 in X we have

$$X_1 + Y_1 = \langle x_0, x_1, x_2, w_0, w_1 \rangle.$$

5.) Let $X \in h_I(U)$. Choose the generators $\{u_1, \dots, u_n\}$ as in 12.1. Then $X = \langle u_1 + t_1, \dots, x_n + t_n \rangle$. From this it follows $\varphi(X) \subseteq \varphi(\langle U \rangle)$. Therefore, the normal independence of the submodules U and V implies the normal independence of X $\in h_I(U)$ and $Y \in h_I(V)$, hence $(X + Y)_1 = X_1 + Y_1$. It follows

$$U + V \subseteq X_1 + Y_1 + T_I \subseteq (X + Y)_1 + T_I.$$

Since

$$rg_I(U + V) = rg_I\ U + rg_I\ V = \dim X + \dim Y = \dim (X + Y)$$

we have $X + Y \in h_I(U + V)$.

Conversely, let $Z \in h_I(U + V)$. Then, because of $U \subseteq Z_1 + T_I$ and $V \subseteq Z_1 + T_I$, there exist $X \in h_I(U)$ and $Y \in h_I(V)$ both contained in Z, hence $X + Y \subseteq Z$. As it was shown above, we have $X + Y \in h_I(U + V)$. The minimality of Z then implies $Z = X + Y$.

Exercises of Chapter 4

1.) Let $a \longmapsto \bar{a}$ be the canonical homomorphism $R \longrightarrow R/M$ where $M = \langle x, y\rangle$ is the maximal ideal of R. We may identify R/M with K. If $a = a_0 + a_1x + a_2y \in R$ then $\bar{a} = a_0$.

Define the residue class space $\bar{T}$ as the 3-dimensional space over K and the residue class homomorphism

$$T \longrightarrow \bar{T}, \quad (a_1, a_2, a_3) \longmapsto (\bar{a}_1, \bar{a}_2, \bar{a}_3)$$

A submodule X of T is a point of H(T) if $X = \langle u\rangle$, $\bar{u} \neq 0$, it is a line if $X = \langle u, v\rangle$, $\bar{u}$ and $\bar{v}$ are linearly independent over K. We can consider T as a 9-dimensional vector space over K. Then the points are 3-dimensional, the lines are 6-dimensional subspaces of T.

Let $u = (0, 0, 1)$ and $v = (x, y, 1)$. The R-submodule of T generated by u and v is the 4-dimensional K-subspace spanned by $(0, 0, 1)$, $(0, 0, x)$, $(0, 0, y)$ and $(x, y, 0)$.

Assume U is a direct summand of T, say

$$T = U + V,$$

containing u and v. Suppose that the first two components of the elements of U are in M. Considering the decompositions

$$(1, 0, 0) = u_1 + v_1, \quad (0, 1, 0) = u_2 + v_2$$

with $u_i \in U$, $v_i \in V$ we have

$$xv_1 = (x, 0, 0), \quad yv_2 = (0, y, 0),$$

hence $(x, y, 0) = xv_1 + yv_2 \in U \cap V$ contrary to the assumption that $U + V$ is a direct sum. Therefore there exists an element $w \in U$ whose first or second component is not in M.

Then u, xu, yu, w, xw, yw, and $(x, y, 0)$ are 7 linearly independent vectors in U. Since U is a direct summand, we obtain $U = T$.

2.) Let $X, Y \in H(T)$. We may assume $\dim X \leq \dim Y$ and further that X is not contained in Y. Then we distinguish the following cases:

(a) dim X = 1, dim Y = 1. The dimension formula yields dim V(X,Y) = 2. This is equivalent to (1).

(b) dim X = 1, dim Y = 2. The dimension formula yields dim V(X,Y) = 3. This is equivalent to (5).

(c) dim X = 2, dim Y = 2. The dimension formula yields dim S(X,Y) = 1 and dim V(X,Y) = 3 or dim S(X,Y) = 0 and dim V(X,Y) = 4. This is equivalent to (2) and (3).

(d) dim X = 2, dim Y = 3. The dimension formula yields dim S(X,Y) = 1. This is equivalent to (4).

(f) dim X = 3, dim Y = 3. The dimension formula yields dim S(X,Y) = 2. This is equivalent to (6).

3.) It is easy to see that there exist subspaces X, Y, and Z of T with $\tilde{X} = \varphi_I(X)$, $\tilde{Y} = \varphi_I(Y)$, $\tilde{Z} = \varphi_I(Z)$ satisfying the following conditions:

There exist normal bases $\{x_1, \ldots, x_k\}$ of X and $\{y_1, \ldots, y_n\}$ of Y such that

$$w_i = y_i - x_i = 0 \quad \text{for } i = 1, \ldots, r$$

$$w_i \notin T_I \quad \text{for } i = r+1, \ldots, t$$

and $\{x_1, \ldots, x_k, y_{t+1}, \ldots, y_n, w_{r+1}, \ldots, w_t\}$ is a normally independent system of generators of $X_1 + Y_1$ and a K-basis of Z. We normalize w_i for $i = r+1, \ldots, t$

$$w_i = a_i w_i', \quad a_i \in K, \quad \|w_i'\| = |1|.$$

Since the valuation is a V-valuation, there exists by 6.4 a non-zero ideal I′ contained in the right ideals $a_i I$, $i = r+1, \ldots, t$.

By 14.2 the union set of $\varphi_I'(X)$ and $\varphi_I'(Y)$ in H_I' contains $\varphi_I'(Z')$ with

$$Z' = \langle x_1 + u_1, \ldots, x_k + u_k, y_{t+1} + u'_{t+1}, \ldots, y_n + u'_n, w_{r+1} + v_{r+1}, \ldots, w_t + v_t \rangle$$

$$= \langle \quad \ldots \quad , w'_{r+1} + a^{-1}_{r+1} v_{r+1}, \ldots, w'_t + a^{-1}_t v_t \rangle$$

with $u_i, u_i', v_i \in T_I'$. Since $I' \subseteq a_i I$ we have $a_i^{-1} v_i \in T_I$, hence $\varphi_I(Z') = \varphi_I(Z)$ by 13.2.

4.) By exercise 4 of chapter 3 we have

$$X_1 + Y_1 = \langle x_0, x_1, x_2, w_0, w_1 \rangle$$

with

$$x_0 = (-1, 1, t, 0, 0) \qquad w_0 = (0, 0, 0, 0, t^3)$$

$$x_1 = (1, t, 0, 0, 0) \qquad w_1 = (0, 0, 0, t^2, 0)$$

$$x_2 = (0, 0, 1, 0, 0)$$

Since the valuation of $K = k(t)$ is discrete, the non-trivial ideals of the valuation ring are

$$I = (t^n), \quad n = 1, 2, \ldots$$

We must distinguish the following cases

a) Let $I = (t)$ or $I = (t^2)$. Then $w_0, w_1 \in T_I$, hence the union and intersection are unique. The first is $\varphi_I(X)$, the second $\varphi_I(Y)$.

b) Let $I = (t^3)$. Then $w_0 \in T_I$, $w_1 \notin T_I$, hence $r = 1$. The members of the union set are now of dimension 4. By 14.2 they are represented as $\varphi_I(Z)$ with $Z = X + \langle w_1 + v_1 \rangle$, $v_1 \in T_I$. Normalizing $w_1 + v_1$, we obtain $Z = X + \langle z \rangle$ with

$$z = (0, 0, 0, 1+a, b), \quad a, b \in t^{-2}(t^3) = (t).$$

By 14.5 the members of the intersection set are of dimension 1 and are described as $\varphi_I(Z)$ with

$$Z = \langle x_0 + u_0 \rangle, \quad u_0 \in I_1 x_1 + T_I$$

where $I_1 = I\|w_1\|^{-1} = (t^3)t^{-2} = (t)$, hence

$$Z = \langle z \rangle, \quad z = (-1+a, 1+b, t, 0, 0), \quad a \in (t), \ b \in (t^2).$$

c) Let $I = (t^n)$, $n > 3$. The union is the whole space, the intersection the null space.

5.) By definition $P_l(I)$ is the maximal ideal of $O_l(I) = \{a \in K \mid aI \subseteq I\}$, hence the set of the non-invertible elements, thus $P_l(I) = \{a \in K \mid aI < I\}$ resp. $P_r(I) = \{a \in K \mid Ia < I\}$.

Exercises of the appendix

1.) Let B_i be the valuation ring of K belonging to U_i and let M_i be the maximal ideal of B_i.

(1) ⇒ (2). Since there are no further φ-convex subgroups between U_i and U_{i+1}, there are no further valuation rings of K between B_i and B_{i+1}. Thus, B_i/M_{i+1} is a maximal locally invariant (i.e. invariant) valuation ring of B_{i+1}/M_{i+1}. Therefore, $(B_i/M_{i+1})\backslash(M_i/M_{i+1})$ is normal in $(B_{i+1}/M_{i+1})\backslash(M_{i+1}/M_{i+1})$. This means that $B_i\backslash M_i$ is normal in $B_{i+1}\backslash M_{i+1}$. Finally, U_i is normal in U_{i+1}.

(2) ⇒ (1). Let $x \in K^x$, $|x| < |1|$ and let M_i be the minimal completely prime ideal of B containing x. $x \in M_i\backslash M_{i+1}$. Since U_i is normal in U_{i+1}, $B_i\backslash M_i$ is normal in $B_{i+1}\backslash M_{i+1}$. By $x \in B_{i+1}\backslash M_{i+1}$, we get $x(B_i\backslash M_i)x^{-1} = B_i\backslash M_i$. It follows $xB_ix^{-1} = B_i$ and $xM_ix^{-1} = M_i$.

2.) a) Let $a_i \in K$ for which $|a_i|_i = \varepsilon_i$ is valid. The cases $B_1 \subseteq B_2$ or $B_2 \subseteq B_1$ are trivial. By 7.3, there exist $z_1, z_2 \in K$ satisfying $|z_1|_1 = |1|_1$, $|z_1|_2 < |1|_2$, $|z_2|_1 < |1|_1$, $|z_2|_2 = |1|_2$. $|z_1|_{1\cdot 2} = |z_2|_{1\cdot 2} = |1|_{1\cdot 2}$. $(|a_1z_1z_2|_1, |a_2z_1z_2|_2, a_1, a_2)$ is compatible. Thus, $|x-a_1|_1 \leq |a_1z_1z_2|_1 < |a_1|_1$ and $|x-a_2|_2 \leq |a_2z_1z_2|_2 < |a_2|_2$ holds for an $x \in K$. We get $|x|_1 = |a_1|_1$ and $|x|_2 = |a_2|_2$.

b) There exists an $x \in K$ satisfying $|x-a_i|_i \leq \varepsilon_i$ ($i = 1,2$). By a), we get a $y \in K$, such that $|y|_i = \varepsilon_i$ holds. If $|x-a_i|_i < \varepsilon_i$, we have $|x+y-a_i|_i = \varepsilon_i$. In the case $|x-a_1|_1 < \varepsilon_1$, $|x-a_2|_2 = \varepsilon_2$ we are ready if $|x+y-a_2|_2 = \varepsilon_2$. Otherwise, $B_1 \not\subseteq B_2$ can be assumed without restrictions. By 7.3, there exists $z \in K$ satisfying $|z|_1 = |1|_1$, $|z|_2 < |1|_2$. It follows $|x+yz-a_i|_i = \varepsilon_i$ ($i = 1,2$).

3.) $|\ |_i'$ denotes the valuation of $B_{1\cdot 2}/M_{1\cdot 2}$ belonging to $B_i/M_{1\cdot 2}$. Let $y_i \in K$ for which $|y_i|_i = \varepsilon_i$ holds. If $|y_1|_{1\cdot 2} = |y_2|_{1\cdot 2} > |a_1-a_2|_{1\cdot 2}$ is valid, then put $x = a_2$. If $|y_1|_{1\cdot 2} = |y_2|_{1\cdot 2} = |a_1-a_2|_{1\cdot 2}$ is valid, then define $z_i := (a_1-a_2)^{-1}y_i$. $z_i \in B_{1\cdot 2}\backslash M_{1\cdot 2}$. By the independence of $|\ |_1'$, $|\ |_2'$ and by 3.1, there exists $y \in B_{1\cdot 2}$, such that

$$|\bar{1}-\bar{y}|_1' < |\bar{z}_1|_1', \quad |\bar{y}|_2' < |\bar{z}_2|_2'$$

holds. Put $x = (a_1-a_2)y + a_2$.

References

ALBRECHT, U., TOERNER, G.

[1] Group rings and generalized valuations. Comm. Algebra 12(18) (1984), 2243 - 2272

BOURBAKI, N.

[1] Commutative Algebra. Addison-Wesley, Reading, Mass. 1970

BRUNGS, H. H.

[1] Rings with a distributive lattice of right ideals. J. Algebra 40 (1976), 392 - 400

BRUNGS, H. H., TOERNER, G.

[1] Extensions of chain rings. Math. Z. 185 (1984), 93 - 104

CHEHATA, C.G.

[1] An algebraical simple ordered group. Proc. Lond. Math. Soc. 2 (1952), 183 - 197

COHN, P. M.

[1] Skew Field Constructions. Cambridge 1977

[2] On extending valuations in division algebras. Studia Sci. Math. Hungar. 16 (1981), 65 - 70

DUBROVIN, N. I.

[1] Chain domains. Moscow Univ. Math. Bull. 36 (1980), 56 - 60

[2] An example of a chain prime ring with nilpotent elements. Math. USSR Sbornik, 48 (1984), no. 2, 437 - 444

ENDLER, O.

[1] Valuation Theory. Springer-Verlag, Berlin/Heidelberg/ New York 1972

GRAETER, J.

[1] Zur Theorie nicht kommutativer Prüferringe. Arch. Math. 41 (1983), 30 - 36

[2] Lokalinvariante Bewertungen. to appear in Math. Z. (1986)

[3] Ueber Bewertungen endlich dimensionaler Divisionsalgebren. Resultate der Math. 7 (1984), 54 - 57

HJELMSLEV, J.

[1] Die natürliche Geometrie. Abh. Math. Sem. Univ. Hamburg 2 (1923), 1 - 36

KAPLANSKY, I.

[1] Topological methods in valuation theory. Duke Math. J. 14 (1947), 527 - 541

KLINGENBERG, W.

[1] Projektive und affine Ebenen mit Nachbarelementen. Math. Z. 60 (1954), 384 - 406

[2] Desarguessche Ebenen mit Nachbarelementen. Abh. Math. Sem. Univ. Hamburg 20 (1955), 97 - 111

[3] Projektive Geometrien mit Homomorphismus. Math. Ann. 132 (1956), 180 - 200

[4] Projektive Geometrie und lineare Algebra über verallgemeinerten Bewertungsringen. Algebraical and Topological Foundations of Geometry. Proc. Coll. Utrecht 1959. Pergamon, Oxford (1962), 99 - 107

KOWALSKY, H.-J., DUERBAUM, H.-J.

[1] Arithmetische Kennzeichnung von Körpertopologien. J. Reine Angew. Math. 191 (1953), 135 - 151

KRULL, W.

[1] Allgemeine Bewertungstheorie. J. Reine Angew. Math. 167 (1932), 160 - 196

MATHIAK, K.

[1] Homomorphismen projektiver Räume und Hjelmslevsche Geometrie. J. Reine Angew. Math. 254 (1972), 42 - 73

[2] Ein Beweis der Dimensionsformel in projektiven Hjelmslevschen Räumen. J. Reine Angew. Math. 256 (1972), 215 - 220

[3] Kennzeichnende Eigenschaften bewerteter Vektorräume. J. Reine Angew. Math. 260 (1973), 127 - 132

[4] Projektive Hjelmslevsche Räume im nicht invarianten Fall. J. Reine Angew. Math. 291 (1977), 182 - 188

[5] Bewertungen nicht kommutativer Körper. J. Algebra 48 (1977), 217 - 235

[6] Zur Bewertungstheorie nicht kommutativer Körper. J. Algebra 73 (1981), 586 - 600

[7] Der Approximationssatz für Bewertungen nicht kommutativer Körper. J. Algebra 76 (1982), 280 - 295

MACHALA, F.

[1] Fundamentalsätze der projektiven Geometrie mit Homomorphismus. Rozpravy Ceskoslovenska Akad. Ved. Rada Mat. a Priv. ved. 90, Sesit 5.

NEUMANN, B. H.

[1] On ordered division rings. Amer. Math. Soc. Trans. 66 (1949), 202 - 252

ORE, O.

[1] Linear equations in non-commutative fields. Ann. of Math. 32 (1931), 463 - 477

[2] Theory of non-commutative polynomials, Ann. of Math. 34 (1933), 480 - 508

RADO, F.

[1] Non-injective collineations on some sets in Desarguesian projective planes and extension of non-commutative valuations. Aequat. Math. 4 (1970),307 - 321

RIBENBOIM, P.

[1] Le theoreme d'approximation pour les valuations de Krull. Math. Z. 68 (1957),1 - 18

[2] Theorie des valuations. Presses de l'Universite de Montreal, Montreal 1968

SCHILLING, O. F. G.

[1] Noncommutative valuations. Bull. Am. Math. Soc. 51 (1945), 297 - 304

[2] The Theory of Valuations. Math. Surveys, IV (1950)

STEPHENSON, W.

[1] Modules whose lattice of submodules is distibutive. Proc. London Math. Soc. (3) 28 (1974), 291 - 310

STONE, A. L.

[1] Nonstandard analysis in topological algebra. in Applications of Model Theory to Algebra, Analysis and Probability. Luxemburg, W. A. J., ed., Holt, Rinchart and Winston, New York 1969

TOERNER, G.

[1] Eine Klassifizierung von Hjelmslev-Ringen und Hjelmslev-Ebenen. Mitt. Math. Sem. Gießen 107 (1974)

VAN DER WAERDEN, B.

[1] Algebra II, Springer-Verlag (1967)

WEBER, H.

[1] Zu einem Problem von H. J. Kowalsky. Abh. Braunschw. Wiss. Gesellschaft 29 (1978), 127 - 134

WIESLAW, W.

[1] Topological Fields, Acta Universitatis Wratislaviensis, Wroslaw 1982

Index